Emergency Incident Media Coverage

EMERGENCY INCIDENT MEDIA COVERAGE

ROBERT S. FLEMING, ED.D.

Fire Engineering®

Disclaimer. The recommendations, advice, descriptions, and the methods in this book are presented solely for educational purposes. The author and publisher assume no liability whatsoever for any loss or damage that results from the use of any of the material in this book. Use of the material in this book is solely at the risk of the user.

PennWell Corporation
1421 South Sheridan Road
Tulsa, Oklahoma 74112-6600 USA

800.752.9764
+1.918.831.9421
sales@pennwell.com
www.FireEngineeringBooks.com
www.pennwellbooks.com
www.pennwell.com

Marketing Manager: Amanda Alvarez
National Account Manager: Cindy J. Huse

Director: Mary McGee
Managing Editor: Marla Patterson
Production Manager: Sheila Brock
Production Editor: Tony Quinn
Cover Designer: Vivian Baldwin

Library of Congress Cataloging-in-Publication Data

Fleming, Robert S.
Emergency incident media coverage / Robert S. Fleming, Ed.D.
pages cm
Includes bibliographical references and index.
ISBN 978-1-59370-310-3
1. Fire departments--Public relations. 2. Fire departments--Management.
3. Emergency medical services--Management. 4. Emergencies--Press coverage.
I. Title.
TH9158.F5934 2012
070.4'4936334--dc23
2013001635

Printed in the United States of America

1 2 3 4 5 17 16 15 14 13

Contents

Preface xi

PART 1

Chapter 1 Media Coverage of Emergency Incidents 3
Stakeholders Have Expectations 5
Emergency Incidents Begin as Local Events 5
Not All Incidents Will Be Routine 6
Things Can Change Quickly 7
Technology Has Changed News Coverage 7
Success in Covering Emergency Incidents 8
Examples of Uninformed Reporting 10
Preparation Is the Key 11
Specialists versus Generalists 12
Covering Emergency Incidents 13

Chapter 2 Emergency Incident Coverage—The Fire Department Perspective 17
A Shared Bias for Action 18
Incident Management Process and Priorities 19
Initial Focus in Incident Management and Operations 21
Reporting of the Incident 23
Getting Information to the Public 25
Enhancing Reporting through Related Stories 26
Community Risk Reduction 28
Achieving Success Together 28

Chapter 3 Emergency Incident Coverage—The Media Perspective 31
The Essence of News Coverage 32
The News Media and Their Expectations 33
Expectations Will Differ by Media Type 35
Technology and News Reporting 40
The Reporter's Needs in Covering an Incident 41

Reporters Have Bad Days Too 42
The Media—Friend or Foe? 43

Chapter 4 The Fire Department Public Information Officer (PIO) 45
Benefits of a Proactive Public Information Approach 46
Roles and Responsibilities of the PIO 48
Role of the PIO at Emergency Incidents 51
Qualifications of a Successful PIO 55
Professional Development of the PIO 58
Serving as PIO at an Incident 60

Chapter 5 The Reporter's Job—Get the Story! 65
Additional Roles in Successful Media Coverage 66
The Reporter's Job 70
How Technology Has Changed News Reporting 72
Elements of Successful News Coverage 73
Preparation for Successful Emergency Incident Reporting 74

Chapter 6 Working Together in Emergency Incident Coverage—The Fire Department and the Media 77
Successful Media Coverage—A Shared Responsibility 78
Developing a Positive Working Relationship 79
Working with the Media: General Principles 82
Working with the Media: Before an Emergency Incident 85
Working with the Media: During an Emergency Incident 86
Working with the Media: After an Emergency Incident 87

Chapter 7 Emergency Incident Coverage in an Age of the Internet and Social Media 91
The Internet and Social Media 92
Changing Paradigms in News Reporting 96
Changing Players in Emergency Incident Coverage 97
Impact of the Internet and Social Media in Breaking News Stories 100

PART 2

Chapter 8 Incident Management 107
Incident Management Priorities 108
Working Together in Media Coverage 110
Incident Management System 111
The Incident Commander within NIMS 113

The Public Information Officer within NIMS 115
The Incident Management Process 118
Developing an Incident Action Plan 118
Resources for Emergency Incident Media Coverage. 121
Learning More about Incident Management 121

Chapter 9 Accidents 125
Typical Incidents. 125
Incident Priorities 127
Operational Overview 127
Safety Considerations 129
Resource Requirements 130
Media Coverage of Accidents 131

Chapter 10 Building Fires 139
Typical Incidents. 140
Incident Priorities 142
Operational Overview 143
Safety Considerations 145
Resource Requirements 146
Media Coverage of Building Fires 147

Chapter 11 Equipment Fires 155
Typical Incidents. 156
Incident Priorities 158
Operational Overview 158
Safety Considerations 161
Resource Requirements 162
Media Coverage of Equipment Fires 163

Chapter 12 Hazardous Materials Fires 169
Typical Incidents. 170
Incident Priorities 172
Operational Overview 172
Safety Considerations 175
Resource Requirements 176
Media Coverage of Hazardous Materials Fires 177

Chapter 13 Outside Fires. 185
Typical Incidents. 185
Incident Priorities 188
Operational Overview 188

Safety Considerations 192
Resource Requirements 194
Media Coverage of Outside Fires 194

Chapter 14 Vehicle Fires 203
Typical Incidents 204
Incident Priorities 206
Operational Overview 206
Safety Considerations 209
Resource Requirements 210
Media Coverage of Vehicle Fires 211

Chapter 15 Hazardous Materials Incidents 217
Typical Incidents 218
Incident Priorities 221
Operational Overview 221
Safety Considerations 223
Resource Requirements 224
Media Coverage of Hazardous Materials Incidents 225

Chapter 16 Infrastructure Emergencies 233
Typical Incidents 234
Incident Priorities 237
Operational Overview 237
Safety Considerations 239
Resource Requirements 241
Media Coverage of Infrastructure Emergencies 241

Chapter 17 Rescues 249
Typical Incidents 250
Incident Priorities 253
Operational Overview 253
Safety Considerations 256
Resource Requirements 257
Media Coverage of Rescues 258

Chapter 18 Searches 267
Typical Incidents 268
Incident Priorities 270
Operational Overview 270
Safety Considerations 273
Resource Requirements 274

Media Coverage of Searches. 275

Chapter 19 Terrorism Incidents 283
Typical Incidents. 285
Incident Priorities 287
Operational Overview 287
Safety Considerations 289
Resource Requirements 291
Media Coverage of Terrorism Incidents 292

Chapter 20 Weather-Related Events and Natural Emergencies 301
Typical Incidents. 302
Incident Priorities 304
Operational Overview 305
Safety Considerations 306
Resource Requirements 309
Media Coverage of Weather-Related Events and Natural Emergencies 309

Chapter 21 Civilian Injuries and Fatalities 319
Civilian Fire Injuries and Fatalities 320
Stakeholder Expectations for Coverage of Civilian Injuries and Fatalities 320
Media Coverage of Civilian Injuries and Fatalities 322

Chapter 22 Firefighter Injuries and Fatalities 329
Firefighter Injuries and Fatalities. 330
An Incident within an Incident—The Rescuer Becomes the Rescued 331
Stakeholder Expectations for Coverage of Firefighter Injuries and Fatalities 333
Media Coverage of Firefighter Injuries and Fatalities. 333

Chapter 23 Arson and Incendiary Fires. 341
Fire Service Roles in Fire Investigation 342
Determining the Origin and Cause of a Fire. 343
Classifying the Cause of a Fire 344
Collecting and Preserving Evidence 344
Media Coverage of Arson and Incendiary Fires 345

Chapter 24 Incidents of National Consequence. 351
Incidents of National Consequence or Interest 352

Managing Incidents of National Consequence or Interest 354
The National Players Take the Stage 355
Ensuring Successful Media Coverage. 356

Chapter 25 Final Thoughts 359

APPENDICES

Appendix A Emergency Service Terminology. 365
Appendix B Emergency Service Acronyms 381
Appendix C Emergency Service Positions. 383
Appendix D Emergency Service Apparatus 387
Appendix E Emergency Service Equipment. 391
Appendix F Governmental Agencies and Private Organizations 395
Appendix G Public Information Officer Sample Position Description . . . 399
Appendix H National Incident Management System Overview. 403

Index 411

Preface

It is virtually impossible to turn on a television news broadcast, tune in to a news radio station, or pick up a newspaper without viewing, hearing, or reading about an emergency incident that is occurring or has occurred within a community, region, or nation. The scope, frequency, and number of emergency incident related news stories have significantly increased in recent years. Coverage of emergency incidents is important to both the media and those who rely on the media for timely, accurate, comprehensive, and professional coverage of emergency incidents and related stories.

It is appropriate that the stakeholders of this news coverage expect the media to get the story right and to report information that is not only accurate, but also timely, comprehensive, and professionally reported. When news coverage is inaccurate, issues can arise between fire and emergency service personnel who are on the scene to resolve the emergency situation and their media counterparts who are there to gather the necessary information to enable them to provide timely reporting on a story of community and perhaps broader interest. These unfortunate issues that challenge the ability of each party to successfully perform their roles and responsibilities can arise from a lack of understanding of the roles, responsibilities, and challenges that each faces. This book was written to provide fire and emergency service personnel and their counterparts in the news media with a resource as they prepare to contribute to enhanced media coverage of emergency incidents and to enhanced accuracy, completeness, and professionalism of media coverage.

A significant challenge that both the news media and fire and emergency services face derives from the fact that we have evolved in a relatively short number of years from a limited number of local and affiliated television stations to an expanded array of news stations, each desiring to be the first to cover the breaking and high-profile stories of the day, which often involve emergency incidents. There is no doubt that we live in an age of instantaneous 24/7 electronic media and news reporting. News radio stations likewise want to be the first to bring breaking stories to their listeners. The limitations of the traditional publication deadlines and schedules for when newspapers "put a story to bed" no longer limit many of these news organizations. They now can disseminate a breaking news story in a timely manner through their websites and online editions and through the increasing use of the various social media that newspaper, radio, and television media outlets now utilize in the interest of providing timely and informative news coverage to those who look to them for this information.

Advances in technology have significantly changed the ways in which the traditional news media—television, radio, and newspapers—gather, report, and disseminate the news, particularly when it relates to breaking stories such as those involving emergency incidents. While these changes certainly reflect an evolutionary process, it would be more accurate to categorize them as revolutionary in terms of their impact. The Internet in particular has greatly enhanced media coverage of emergency incidents and provides a convenient forum for interested parties to become informed about these incidents.

While the Internet has been around for a number of years, it is only in recent years that its potential for information dissemination has been fully understood and its uses appreciated. The advent and explosive popularity and growth of social media has taken emergency incident news coverage to the next level. In a sense, social media prompt us to rethink the various entities that comprise what we have traditionally referred to as the "news media." The Internet and social media have provided platforms for anyone who has the capability to capture and disseminate digital images or video from an incident scene to become a "reporter," and an increasing number of individuals have done so already and will likely do so in the future. This certainly changes the manner in which a growing number of individuals get information regarding the occurrence and developments of emergency incidents.

The enhanced potential for timely, practically instantaneous, provision of commentary or images from an emergency incident scene is a reality in our technology-enabled society. A growing number of fire and emergency service organizations, typically through a public information officer (PIO), are likewise utilizing the Internet and social media to communicate directly with the public and the news media. A major advantage of so doing is that these organizations now have a forum that enables them to communicate an unfiltered message directly to its intended recipients. While the Internet and social media can certainly be used by fire and emergency service organizations to their advantage, these communication tools also present a potential and often significant downside in terms of the organization being portrayed in a negative light. Digital images or video footage of a fire or emergency service organization's perceived inaction or inappropriate actions can quickly bring the organization's capabilities and standing in the community into question, thus potentially undermining public and governmental support at a point in time when budget cuts, reductions in force, consolidation, and regionalization are topics of interest and discussion in many jurisdictions.

The unfortunate reality is that while the advent, accessibility, and popularity of the Internet and social media can be instrumental to the reputation and standing of the contemporary fire or emergency service organization, the fact is that the individuals initiating the related incident information can frame or present that information from their perspective, which can significantly influence how the recipient of a message decodes or interprets that message. Such "news coverage" can bring into question the strategies, tactics, and actions of a fire department that did everything correctly, such as allowing a warehouse containing extremely hazardous materials to burn while protecting exposure buildings rather than applying water to the involved building with a resulting negative environmental impact. It has become commonplace for the news media to comment through a voice over on video that they either shot or that they secured from individuals who were on the scene before their arrival, and in some cases before the arrival of the fire department. Through working collaboratively with the media at an incident scene or issuing independent communications through the use of the Internet and social media, a PIO has an essential role and opportunity to preserve the reputation of his or her organization within the community that it serves, as well as in the significantly larger technology-enhanced media footprint that exists today.

Enhancing the accuracy, comprehensiveness, professionalism, and timeliness of media coverage involves a shared responsibility between fire and emergency service personnel and their counterparts in the news media. Only through working together can both parties succeed in terms of the news coverage of emergency incidents. This book was written to provide members of both the media and fire and emergency services with an understanding of the processes, dynamics, and challenges of incident management and successful media coverage of emergency incidents. Through understanding the roles, responsibilities, perspectives, and challenges that each party faces, both will be better prepared to collaborate in a manner that enhances the accuracy, comprehensiveness, professionalism, and timeliness of the news coverage that is provided to interested stakeholders. The first part of this book considers these issues, examining media coverage of emergency incidents, the fire department and media perspectives of emergency incident coverage, the roles and responsibilities of the PIO and reporters in emergency incident reporting, and working together in coverage of emergency incidents. In addition to addressing the impact and implications of the Internet and social media throughout the various chapters in part 1, a dedicated chapter on emergency incident coverage in an age of the Internet and social media is included.

The second part of the book is designed to contribute to the preparation of both parties to emergency incident media coverage in advance of the occurrence of an emergency incident, as well as to serve as a useful reference during an incident. In addition to a chapter that provides a necessary, and often lacking, understanding of incident management for members of the news media, 16 incident-specific chapters are included that provide guidance to reporters and other news media professionals regarding the key information and issues that should be included in comprehensive reporting on each type of incident. This correspondingly alerts fire and emergency service personnel, whether serving as designated departmental PIOs or assuming these responsibilities at a given incident, about the information that they should assemble prior to being interviewed by reporters or conducting a press conference. Information is provided in these various chapters on relevant aspects of informed news coverage of each category of emergency incident.

The appendices are intended to serve as a reference for the media in terms of understanding fire and emergency service terminology, acronyms, positions, apparatus, equipment, and agencies, as well as the roles and responsibilities of the PIO within the National Incident Management System (NIMS) that is in use throughout the United States.

Through communication, cooperation, and collaboration, fire and emergency service personnel and media representatives can work together in a mutually beneficial manner that will enhance their ability to ensure that the news coverage of emergency incidents is accurate, comprehensive, professional, and timely. Successful media coverage of emergency incidents will not always be easy, but it is certainly achievable. An essential component in the pilgrimage to enhance media coverage in your community is obviously the development of the crucial relationships and mutual respect between fire and emergency service personnel and their counterparts in the news media. Relationships built in advance of an emergency incident are instrumental in the successful media coverage of each and every incident. Together, both responders and news media personnel can ensure that this news coverage not only meets, but also exceeds, the expectations of those who rely on it.

PART 1

This book's intention is to assist fire and emergency service professionals and their counterparts in the news media in their pilgrimage of preparing to deliver news coverage of emergency incidents that fully meets and ideally exceeds stakeholder expectations in terms of timely, accurate, comprehensive, and professional reporting. It is written and organized to serve as a preparatory tool and professional resource for both audiences—fire and emergency services personnel and the news media.

The book is organized into three logical and interrelated components. The seven chapters in part 1 consider the essentials of emergency incident media coverage, including the expectations that stakeholders have for this coverage and how fire and emergency services personnel can successfully cooperate and collaborate with their news media counterparts to enhance the media coverage of emergency incidents. Part 2 consists of 18 chapters that consider media coverage of various types of emergency incidents. A set of eight appendices are included in the interest of providing useful reference information to the parties to effective media coverage of emergency incidents.

The first chapter provides an overview of media coverage of emergency incidents. Chapters 2 and 3 examine the goals and challenges of emergency incident media coverage from two essential perspectives—that of the fire department and that of the news media. The roles and responsibilities of both the fire department and the media are respectively considered in chapter 4, which examines the role and responsibilities of the fire department's public information officer (PIO), and chapter 5, which discusses the role of a news reporter in the coverage of an emergency incident. Chapter 6 integrates the material discussed in the previous five chapters in terms of developing, nurturing, and creating the professional working relationship between fire and emergency services personnel and the media necessary to enhance the quality of news reporting in terms of timeliness, accuracy, comprehensiveness, and professionalism.

In addition to discussing the many traditional challenges of working together in successful news coverage of emergency incidents, the aforementioned chapters also consider the new challenges that derive from the advent, commercialization, and availability of new technologies. The Internet and social media present both opportunities and threats in terms of media coverage of emergency incidents. While these issues are considered as appropriate in the first six chapters, the significance of these new technologies to the successful media coverage of emergency incidents makes it appropriate that the final chapter in part 1 be devoted to emergency incident coverage in an age of the Internet and social media.

Chapter 1

Media Coverage of Emergency Incidents

Chapter Objectives

- Discuss the expectations that stakeholders have for emergency incident media coverage.
- Discuss how emergency incidents begin as local events.
- Discuss how things can change quickly during an emergency incident.
- Discuss how technology has changed news coverage of emergency incidents.
- Identify the elements of successful emergency incident media coverage.
- Identify examples of uninformed reporting of emergency incidents.
- Discuss the importance of preparation in successful media coverage of emergency incidents.

Every day fire and emergency service organizations across the nation and around the world are called upon to respond to emergency incidents within the communities they serve. While their call volume and the nature and corresponding challenges of the incidents to which these organizations respond will vary from day to day and from jurisdiction to jurisdiction, the reality is that the stakeholders of fire departments and other emergency service organizations rely on these organizations to respond when an emergency situation or event occurs and to fulfill their roles and responsibilities in a manner that leads to the effective, efficient, and safe resolution of the emergency situation.

The National Fire Protection Association (NFPA) reported that in 2011 there were 30,145 fire departments in the United States with 2,550 considered "all career," in that all of their personnel were paid. The remaining fire departments were classified as "mostly career" (1,865), "mostly volunteer" (5,530), and "all volunteer" (20,200).[1] Regardless of the staffing arrangement utilized, the members of each community have a set of expectations for the fire department and other emergency service organizations that serve them.

Just as the fire department and other relevant emergency service organizations are alerted through dispatch processes and protocols upon the receipt of a call for assistance by the jurisdiction's 9-1-1 center, the members of the local news media become aware of the emergency situation or event that has occurred within a community that is part of their "beat" or coverage area. The same stakeholders that have expectations for their fire department and other emergency service organizations likewise have expectations that their local media will assign the story to media professionals, including reporters, photojournalists, editors, and news anchors that have the knowledge and skills to provide coverage of the emergency incident in a timely, accurate, comprehensive, and professional manner.

At present there are about 1,400 daily and 6,200 weekly newspapers in the United States.[2] Approximately 10,000 commercial and 2,500 non-commercial radio stations exist.[3] Television media coverage is provided by more than 50 nationwide broadcasting networks, as well as numerous affiliate and independent broadcast stations around the nation.[4]

This book has been written as a resource guide for both the media and fire and emergency service organizations. It is based on an underlying assumption that news coverage of emergency incidents that is timely, accurate, comprehensive, and professional represents a realistic stakeholder expectation for the news media covering breaking news stories within their community, just as these same stakeholders have a right to expect that their fire department and related emergency service organizations will respond in a timely manner and perform their roles and responsibilities in an effective, efficient, and safe manner that leads to the resolution of the emergency situation to which they were dispatched.

Stakeholders Have Expectations

Meeting and exceeding the reasonable expectations that have been introduced above and will be further considered in the early chapters of this book, whether as a fire department member or as a media professional, will be based on necessary and appropriate preparation before an incident occurs. This preparation on the part of the fire department will include ensuring that the necessary highly trained personnel, as well as functional apparatus and equipment, are in readiness to respond at the time an emergency occurs. Success in fulfilling the fire department's mission and meeting and, where possible, exceeding reasonable stakeholder expectations demands that fire department personnel have the requisite knowledge, skills, and attitudes to enact their responsibilities successfully.

The same is true of media professionals, whether employed by or affiliated with local or national media organizations. These reporters, photojournalists, news anchors, assignment editors, directors, and producers must likewise have the necessary knowledge, skills, and attitudes to excel in their coverage of each and every breaking news story, including the ever-present and diverse stories that involve emergency incidents and their impact and consequences within a community, and often a much larger geographic area, including incidents that, while occurring locally, by their nature, magnitude, or scope have the potential of becoming incidents of national consequence.

Emergency Incidents Begin as Local Events

The reality is that, regardless of the type of emergency incident, the initial response will involve local resources. The fire department and other local emergency responders will constitute the initial dispatch and response to an incident. As appropriate, the response of additional local, regional, state, and even national resources may be triggered based on the nature of the incident and its resource needs.

An incident of national consequence, such as the terrorist attacks of September 11, 2001, will obviously result in a sizable and comprehensive response of fire department and other emergency organization and agency resources through automatic aid dispatched by protocol based on the nature of an incident and mutual aid wherein an incident commander requests the additional resources deemed necessary to successfully mitigate an emergency situation or event. Illustrative of this would be the response of numerous Federal Emergency Management Agency (FEMA) urban search and rescue (USAR) teams to both the World Trade Center and to the Pentagon on that tragic day in our nation's history.

Likewise, the initial media response to an emergency incident will be the local media, including reporters from newspapers, radio, and television media organizations. The geographic locale in which the incident takes place will have a bearing on the scope of the initial media response and coverage. The incident scene may be within a major metropolitan market, such as New York City, covered by national media organizations, or in a rural area such as Shanksville, Pennsylvania, where the fourth plane of the September 11 terrorist attacks was brought down in a rural field in an area covered by local media and fairly distant from network affiliates and other national news organizations.

Not All Incidents Will Be Routine

While the great majority of emergency incidents that a fire department or local news organization will respond to may be fairly routine incidents, limited in their nature and scope, it is essential that the emergency responders and media representatives who respond to these incidents ensure that the expectations of the communities that they serve are fully met and, where possible, exceeded in terms of a professional response by the fire department and other emergency service organizations and agencies leading to the effective, efficient, and safe resolution of the emergency situation, and the timely, accurate, comprehensive, and professional coverage of the story by the news media.

What a given day holds for a fire department or other emergency service organization and for the media that provide news coverage to those who live in, work in, or travel to or through a community, cannot be predicted with any degree of certainty. While the typical incidents to which the fire department and the media respond will be both routine in nature and local in scope and consequence, when both the fire department and media least expect it, a major regional incident, such as a large fire in an apartment complex, may occur, presenting challenges to both the fire department and to their counterparts in the news media. It is also possible that the event of the day, while occurring at the local level, may very quickly escalate to an incident of national consequence and interest.

Things Can Change Quickly

Those involved in the initial fire department and media response in Shanksville, Pennsylvania, on September 11, 2001, can certainly attest to this reality and the various challenges associated with the escalation of an emergency incident and the corresponding breaking news story from a local incident to a national incident. Experience has shown that both fire department and media representatives can quickly find themselves in front of the national media and that the challenges of dealing with the media can at times be just as daunting as those associated with resolving the emergency situation or event to which they were dispatched and responded.

The news coverage of high-profile incidents has frequently launched the careers of local reporters and local fire department officers. Likewise, being unprepared to contribute to the timely, accurate, comprehensive, and professional news coverage of an emergency incident, regardless of nature or scope, can present the reporter or fire officer in an undesirable light, particularly in the age of instantaneous 24/7 electronic media, potentially derailing what had previously been a promising fire service or media career.

Technology Has Changed News Coverage

There is no debate that advances in technology and the resulting commercialization and application of new technologies have revolutionized the contemporary world. Through the Internet and various social media platforms, anyone with a fairly inexpensive digital still or video camera, or a modern cell phone, has the potential of immediately becoming a news reporter and disseminating their coverage of a breaking news event to a limited number of friends or to a much larger audience through such social media as YouTube or Facebook. In a sense, these enhanced capabilities of an individual, regardless of knowledge or experience, to capture video images and instantaneously share them along with commentary with interested parties can prove a major challenge to both fire and emergency service organizations and to the traditional news media in that while such nonprofessional coverage may correspond to a stakeholder's interest in timely information on breaking news stories such as emergency incidents, it often constitutes reporting that is inaccurate, misleading, or unprofessional.

These individuals constitute a growing cadre of what could be considered independent social media reporters. They often arrive at an emergency incident early, as a result of being in the right place at the right time, and begin to capture the evolving incident before the arrival of the traditional news media, and in some cases the fire department. It is not unusual for the news media to seek out their video footage of the early stages and developments of an incident prior to their arrival. There has also been an increasing trend of television stations encouraging their viewers to capture and send in such video footage for incorporation into their news broadcasts or in their supporting websites. When you think about it, technology has significantly changed the nature of contemporary news coverage and perhaps even redefined the term "news media."

Success in Covering Emergency Incidents

Success, and often survival, in the media coverage of emergency incidents, whether as a fire department representative such as an incident commander or a public information officer (PIO) or as a news reporter or other media representative, will be greatly enhanced through recognizing that we are all in this together. In reality, the contemporary fire department is also a key stakeholder of the news coverage of incidents to which it responds. The fire department, just like the other stakeholders of this news coverage, has a vested interest in ensuring that the media cover the emergency incidents to which it responds in a timely, accurate, comprehensive, and professional manner.

An integral aspect of the expectations of contemporary fire and emergency service organizations with respect to media coverage is that all coverage of the organization operating on an incident scene, as well as other stories related to the organization, be portrayed in a manner that does not inappropriately diminish the organization's standing and reputation in the community that it serves and counts on for support. Reputation management is thus a growing concern and challenge that contemporary fire and emergency service organizations must be prepared to address professionally in a timely manner, often while the emergency scene is still active. All too frequently well-respected fire and emergency service organizations now find their reputation, image, and credibility being compromised as a result of the television airing or Internet posting of digital still or video images captured by not only the traditional news media, but also by the public, and increasingly by first responders operating at the incident scene. The reputations of these organizations and their personnel can either be instantly enhanced or diminished in the social media age in which we live and work (fig. 1–2). That said, there will be times when such coverage depicts aspects of a fire department's operations that should be addressed and improved.

Fig. 1–2. Reputation management is essential in the age of social media.

Successful media coverage of emergency incidents that meets and ideally exceeds the community's expectations for this coverage requires the development of a collaborative working relationship between fire departments and other emergency service organizations and their counterparts in the media whose responsibilities will include covering stories regarding emergency incidents. This relationship, which must be nurtured based on mutual respect and professionalism, should be developed before an incident occurs and invoked throughout the successful management of the incident by the fire department and coverage of the story by the media.

Examples of Uninformed Reporting

Review of the news coverage of emergency incidents frequently reveals that, while the media may meet the expectation of providing incident information in a timely manner, the expectations that the coverage be accurate and comprehensive are often not met. This failure to meet stakeholder expectations is typically a result of a lack of knowledge regarding the nature of the various types of emergency incidents and emergency incident scene operations. Examples of frequently inaccurate, uninformed, and misleading coverage include the following cases:

- Reporters frequently report that firefighters are wearing "oxygen tanks" as they enter burning structures. In reality, the self-contained breathing apparatus (SCBA) that firefighters don when entering hazardous atmospheres utilize normal breathing air, rather than pure oxygen.
- The media coverage for large gas line ruptures involving fire is sometimes critical of firefighters and their departments for not extinguishing the fire in a timely manner. To do so before shutting off the gas supply line would create an even greater hazard in that the leaking gas would be highly explosive and susceptible to rapid combustion and explosion if contact was made with an ignition source.
- In cases of hazardous materials leaks, fire departments and emergency response agencies have often been inappropriately

criticized for not promptly evacuating nearby residents. While there are times that rapid evacuation makes sense, there are times that sheltering in place may be more advisable for a number of life safety and logistical reasons.

- Media sources often question why firefighting personnel decide not to attack fires aggressively by applying large quantities of water. The nature of the fuel load, including the presence of certain hazardous materials, may dictate that the prudent decision is either not to attempt to extinguish the fire and/or to protect exposures.

Preparation Is the Key

As with most things, knowledge and skills enable an individual to professionally fulfill his or her responsibilities, whether as a representative of the media or as a fire department member. While both fire department personnel and their counterparts in the news media are typically well prepared to perform their fundamental roles and responsibilities such as fighting a fire or providing the corresponding basic news coverage of that incident, real success in media coverage of emergency incidents will only come when both parties understand each other's roles, responsibilities, and priorities on an emergency incident scene. They should understand how they can collaborate to ensure the effective, efficient, and safe resolution of the incident and the corresponding timely, accurate, comprehensive, and professional coverage of the related news story.

The intent of this initial chapter is to introduce the reader to the issues associated with media coverage of emergency incidents. The six chapters that follow have been designed to further explore the roles and responsibilities of fire department members and media representatives with respect to an emergency incident, particularly during the mission-critical early minutes and hours of an incident during which the fire department's success in managing the incident and the media's success in covering it are often determined. Chapter 2 considers emergency incident media coverage from the perspective of the fire department, with chapter 3 considering the same incident coverage from the perspective of the news media. These two chapters are thus designed to enable both parties to the successful news coverage of an emergency incident to develop an understanding of both their challenges, and those of their counterparts.

Specialists versus Generalists

The work of any organization, whether a fire department or a television station, is performed by individuals. As an organization grows in size, in terms of number of personnel, it has the opportunity to divide up the work to be done and assign this work to appropriate individuals, through labor specialization, thus allowing the organization to utilize the most qualified individual to handle a given role and its accompanying responsibilities. An example of this in a fire department would be designating an individual with the necessary knowledge, skills, attitude, and willingness to serve as the department's PIO.

Corresponding examples in the news media would involve a television station or a newspaper designating an individual with the proper qualifications and interest to serve as an emergency services reporter. While this is a growing trend in many large media organizations and can significantly enhance the coverage of these news stories, it is often a luxury that small and medium-sized news organizations cannot afford. Thus it is not unusual for assignment editors to assign anyone from the staff who is available and in close proximity to handle breaking news stories related to emergency incidents.

Before assigning a reporter to cover a reported emergency incident, the news organization will make a conscious decision regarding its interest in covering the particular story. Factors considered in making this decision will obviously include the number, scope, and significance of other stories being covered, as well as the availability of reporters and photojournalists to assign to coverage of a particular emergency incident. These decisions are based on the perceived news value or newsworthiness of a potential story. The type of incident does not always drive these decisions. While a major fire at an apartment complex might typically result in extensive news coverage from the incident scene on a slow news day, on a busy day with many breaking stories where the television crews arrived fairly late in the incident, the coverage might be limited to a television news anchor talking over a brief video of the aftermath of the fire. A seemingly less significant fire in a single family residence where there was limited fire damage and no injuries or fatalities might receive extensive coverage if it was the home of a well-known political figure or an entertainment or sports celebrity. The receipt of digital video footage from a viewer that was shot early during a vehicle or structure fire that vividly showed the fire or that the fire department experienced an issue such as establishing a water supply to fight the fire as a result of a defective fire hydrant could also result in the decision by a television station to air the video, especially if it is something that their news media competitors did not have. Last, but certainly not least, on a daily basis there will be many smaller and less significant incidents that the media will pass on covering.

In addition to being written to prepare members of the fire service and the news media who serve in dedicated roles with respect to the media coverage of emergency incidents, such as those discussed above, this book is also intended to provide media representatives or fire department personnel who on a given day at a particular incident may find themselves in a position to contribute to the timely, accurate, comprehensive, and professional news coverage of that incident as a member of the fire service or the news media the knowledge, insights, and tools to succeed. Chapter 4 is thus dedicated to discussing the role and responsibilities of a fire department's PIO in the successful news coverage of an emergency incident, while chapter 5 examines the issues and challenges facing a reporter in getting and reporting the story of an emergency incident. Chapter 6 appropriately focuses on how the fire department and the media can work together to ensure successful media coverage of emergency incidents that meets the expectations of all stakeholders, including both the fire department and the media. The final chapter in part 1 further considers the challenges of emergency incident coverage in an age of the Internet and social media.

Covering Emergency Incidents

While timely reporting on emergency incidents is obviously a primary expectation that the various stakeholders of the information age have for emergency incident media coverage, these same stakeholders have a right to expect that this news coverage will also be accurate, comprehensive, and professional. The second part of this book is designed to be a preparation tool and resource guide for covering the typical incidents to which a fire department will be dispatched and respond, and consequently the news media may have an interest in or responsibility to cover. The various chapters in this part of the book are organized into general categories based on emergency incident type.

Review of the titles of these chapters will reveal that the traditional services provided by fire departments have evolved from the early days when a fire department's role and the services that it provided were typically limited to fire suppression. Over the years these services have evolved to the point where most contemporary fire departments respond to a diverse array of fire, rescue, hazardous materials, and emergency medical incidents. The five chapters that relate to emergency incidents involving fire consider media coverage of building fires, equipment fires, vehicle fires, outside fires, and fires involving hazardous materials.

Additional chapters examine the expanded mission and corresponding services of the contemporary fire department, including media coverage of accidents, hazardous materials incidents, infrastructure emergencies, rescues, searches, terrorism incidents, and weather-related emergencies and natural disasters. Two additional chapters consider news reporting on the tragic outcomes of emergency situations in terms of civilian and/or firefighter injuries and fatalities. A chapter is devoted to media coverage of arson and incendiary fires. A final chapter considers media coverage of incidents of national consequence.

The material in the various chapters in the second part of this book is presented in a manner designed to contribute to the preparation of fire department and media personnel before an incident in terms of enhancing their knowledge regarding particular types of emergency incidents and the information that will contribute to the successful media coverage of these incidents. Operational information is provided with respect to each type of incident, including the nature of the incident; priorities in managing the incident; an operational overview of the strategies and tactics required to resolve the incident; potential issues associated with the incident; safety considerations in managing the incident; typical responding agencies; response personnel, apparatus, and equipment; and relevant terminology. The discussion of this information is designed to give media representatives awareness and understanding, while alerting fire department personnel to some of the things they should be prepared to discuss with the media.

A significant component of the discussion in each of the chapters related to media coverage of emergency incidents is designed to prepare representatives of the news media to successfully gather and report relevant aspects of a given story. This information is also designed to give fire department representatives who will work with the media guidance on the issues that they should be prepared to address in a media interview, as well as some of the questions they can expect to be asked. Each of these chapters considers story sources, background, and development, as well as questions to ask and available informational resources. Each chapter also suggests related sidebar stories that a reporter could consider with respect to an emergency incident.

Each of these chapters incorporates incident guides designed to be used while covering that type of emergency incident. These guides have been prepared to serve as job aids to enable those fire department and media personnel involved in the news coverage of an emergency incident to succeed in ensuring that the coverage meets the expectations of the various stakeholder groups, including their own organizations, in terms of its timeliness, accuracy, comprehensiveness, and professionalism. Such job aids have demonstrated their value in many professions and settings, such as in the management of major emergency incidents, including terrorist events.

An additional resource is provided through appendices that include essential information on emergency service terminology, acronyms, positions, apparatus, and equipment, as well as the National Incident Management System (NIMS) and the federal role at certain emergency incidents. The appendices also include a sample PIO position description.

In summary, this book is intended to be a valued-added resource to enable fire and emergency service organization members and the media to prepare prior to an emergency incident to ensure that the reporting related to that incident fully meets and, where possible, exceeds the expectations of newspaper readers, radio listeners, and television viewers, as well as those of other stakeholders including elected and appointed officials, the fire department and its members, and the media organization. Through thorough preparation, both fire service and media personnel can be ready to deliver the timely, accurate, comprehensive, and professional coverage of emergency incidents that those within the communities that they serve expect and deserve.

Chapter Questions

1. Relate and explain the expectations that stakeholders have for the media coverage of emergency incidents.
2. Discuss how emergency incidents that begin as local events can escalate to events of national consequence or interest.
3. Discuss the dynamic nature of emergency incidents.
4. Discuss how the news coverage of emergency incidents has changed with advances in technology.
5. Relate and explain the elements of successful media coverage of emergency incidents.
6. Provide several examples of uninformed reporting that can prove inaccurate, confusing, or misleading.
7. Discuss the role of preparation in the successful media coverage of emergency incidents.

Notes

1 National Fire Protection Association. 2012. *U.S. Fire Department Profile*. Quincy, MA: National Fire Protection Association.

2 Neuharth, Al. 2008. "On 225th Birthday, Newspapers Dying?" *USA Today*. http://usatoday30.usatoday.com/printedition/news/20080523/al23.art.htm.

3 Farrish, B. n.d. "The Number of Radio Stations in the U.S." Music Biz Academy.com. www.musicbizacademy.com/articles/radio/stations.htm.

4 Wikipedia. 2013. "List of United States Over-the-Air Television Networks." http://en.wikipedia.org/wiki/List_of_United_States_over-the-air_television_networks.

Chapter 2

Emergency Incident Coverage—The Fire Department Perspective

Chapter Objectives

- Discuss how both media and emergency service organizations share a bias for action.
- Discuss the process and priorities utilized in managing an emergency incident.
- Discuss the initial focus in incident management and operations.
- Discuss the elements of successful media coverage of emergency incidents.
- Discuss the importance of disseminating information to the public during emergency incidents.
- Discuss how related sidebar stories can contribute to enhanced reporting of an emergency incident.
- Discuss the concept of community risk reduction.
- Discuss how representatives of the media and emergency services can work together to enhance incident management and emergency incident media coverage.

Among the various stakeholder groups with an interest in timely, accurate, comprehensive, professional media coverage of emergency incidents, there are two primary groups that stand out—fire and emergency service organizations and the news media. Together these groups have both an interest in and a responsibility for ensuring the responsible reporting of emergency incidents, whether of local, regional, national, or international interest. This chapter and the one that follows are devoted to developing an understanding of the perspectives of each group with respect to emergency incident coverage.

The typical expectations that stakeholders have for their fire department, as introduced in the previous chapter, provide a logical starting point for our consideration of the fire department perspective of coverage of emergency incidents by the news media. These stakeholders expect the fire department and other emergency service organizations that serve them to be prepared to meet all of their responsibilities, regardless of the nature or size of an emergency situation, in an effective, efficient, and safe manner. Effectiveness considers whether the response organization(s) are successful in accomplishing what they set out to do, such as bringing a fire under control and extinguishing it. Efficiency further considers the resources utilized in doing so. Last, but most important of all, is that the fire department and other response organizations fulfill their responsibilities in a manner focused on the life safety of response personnel and the public.

A Shared Bias for Action

A primary stakeholder expectation for fire and emergency services is that upon receipt of a request for emergency assistance, these organizations will be dispatched, respond, arrive on the incident scene, and resolve the given emergency situation in a timely manner. The same pressure in terms of getting a news story and sharing it quickly also drives the motives and actions of the news media. An obvious objective of the news media representatives covering an incident is to "scoop the story," beating the competing media organizations to coverage so as to have exclusive coverage of an event that other media organizations do not have. This can often result in a dynamic that could be called a "rush to report."

While most contemporary fire and emergency service organizations and their officers fully understand the desire of the media and their viewers, listeners, or readers to have the news instantaneously, they also have their priorities correct in terms of the need to promptly address the emergency situation and then to work with the media to ensure that their coverage is timely, accurate, comprehensive, and professional.

Were a fire department and its personnel to become distracted in the early minutes of an emergency incident during which critical decisions must be made and actions taken, it would be disregarding its mission and responsibilities to the community that it serves. Thus, fire departments must ensure that the media understand the mission-critical nature of the early stages of many incidents and that the fire department will likely not be able to devote much attention to media coverage of the incident during this time in terms of making personnel available for interviews, particularly those in command of the incident.

The media's coverage of the early minutes of an incident can be enhanced through this understanding and by knowing and being able to report the processes through which incident commanders manage incidents and the various strategies and tactics typically utilized in managing different types of emergency incidents. This understanding will be provided through the various chapters in the second part of this book as they discuss the typical types of emergency incidents to which fire departments and other emergency service organizations and their counterparts in the news media will respond.

Incident Management Process and Priorities

For many years, fire and emergency service organizations have used a common system in managing incidents called the Incident Command System (ICS). The U.S. Department of Homeland Security (DHS) further enhanced this command system, incorporating it into the National Incident Management System (NIMS) that was introduced after the events of September 11, 2001. Emergency response personnel from all disciplines have been trained and are prepared to implement NIMS at major emergency incidents.

A starting point in understanding the management of any emergency incident is recognition of the priorities in managing an emergency incident. The first priority is always life safety, with the second priority of incident stabilization, and the third priority of property conservation. These priorities should drive every decision and action on an emergency incident scene. An essential element of successful media coverage of these incidents is that media representatives understand and can report on these priorities as they cover a developing story.

Under an incident management system, a qualified individual is designated as the incident commander. This person is responsible for the overall management of the incident. In certain incidents a unified command may be utilized wherein the command responsibilities are performed by more than one individual—for example, by the fire chief and police chief together. More information about the various responsibilities of the incident commander is provided in chapter 4.

The successful management and resolution of an emergency situation begins with the incident commander performing a situational evaluation involving what has happened and what is presently happening and forecasting what is likely to happen next. Based on this situational evaluation, referred to as size-up, the incident commander formulates strategic goals and tactical objectives for the incident based on the size-up and in accordance with the three established priorities of incident management. Strategic goals represent desired outcomes, while tactical objectives involve the identification of the actions or tactics that will be necessary to successfully resolve a given emergency situation. Last, the necessary resources in terms of personnel, apparatus, and equipment must be assigned to perform each of the necessary tactics. It should be recognized that size-up should continue throughout the management of an incident in terms of tracking the successful enactment of tactics and their contribution to the attainment of the strategic goals that have been established for the incident (fig. 2–1).

The management of an incident involving a fire in a single-family residence serves to illustrate this process. As with all incidents, the priorities of managing this incident in order of importance would be life safety, incident stabilization, and property conservation. The strategic goals could be to rescue all occupants from the residence and to not injure any firefighters while so doing. The tactics necessary to accomplish these strategic goals would typically include: establishing a water supply, extinguishing the fire through the use of hoselines, laddering the building, performing search and rescue, ventilating the building, and controlling the utilities. Each of these tasks would be assigned to appropriate emergency response personnel and units.

Fig. 2–1. Incident commander directing an incident. (Courtesy of Tony Barbato.)

Initial Focus in Incident Management and Operations

For years practitioners and researchers in emergency medicine have recognized what they refer to as "the golden hour." This short time frame after the occurrence of a medical emergency or traumatic injury has often been found to determine success in terms of patient outcomes, with those patients who receive necessary medical attention at an appropriate medical facility within the first hour after the occurrence of an injury or onset of a medical emergency routinely faring much better in terms of survival and successful recovery.

A similar reality exists with respect to managing an emergency incident and providing news coverage of that incident. There are critical decisions that the incident commander and other officers on an incident scene must attend to and tactics that need to be assigned and enacted in a fairly short period of time within the early minutes and sometimes first hour of an incident. What happens during this time period, similar to the golden hour in emergency medicine, can make the difference between a successful and an unsuccessful outcome.

It is therefore imperative that all media representatives understand and respect the fact that the fire department and other response organizations, and their officers, have things that they must do in the early stages of an incident that will require their time and talents and thus will limit their availability to meet with the news media to discuss or be interviewed regarding the incident. That being said, it is important to restate that the top priority in managing any incident is life safety, and that ensuring the safety of media representatives will obviously be an integral aspect within life safety on an incident scene.

Accordingly, the fire department and other response agencies cannot expect the news media to hold off on initiating coverage of the incident until someone from fire and emergency services is available to talk to them. Thus, reporters will be expected to begin their coverage, with commentary often initially from a studio and later from the incident scene, in a timely manner and to incorporate into their reporting interviews with fire and emergency services personnel as they become available.

Ideally, at an appropriate point in the incident, when a public information officer (PIO) or the incident commander can be made available to meet with the media, a press conference and/or individual interviews can be conducted. The fire department will expect reporters to talk to those knowledgeable and informed fire department representatives authorized and trained to work with the media, rather than try to get interviews from various emergency responders such as firefighters or emergency medical services personnel. The reasonableness of this expectation is further explored in chapter 3, with a discussion of the desired working relationship between a departmental PIO and the news media in chapter 4.

Reporting of the Incident

As previously discussed, television viewers, radio listeners, and newspaper readers rely on the news media, through the assistance of fire and emergency service organizations, to accurately report on emergency incidents. This is especially important in cases where there are civilian and/or firefighter injuries and/or fatalities, when inaccurate reporting can contribute to the stress levels of interested individuals, including loved ones and significant others. Situations where parties are potentially in harm's way and decisions have been made to either shelter in place or to evacuate must be properly reported by the news media based on the provision of accurate information by fire departments and emergency response organizations or appropriate elected or appointed public officials. Responsible reporting that is timely, accurate, comprehensive, and professional must thus become the norm in covering emergency incidents.

The mission of contemporary fire departments and the services that they provide, coupled with the 24/7 news cycles and technological capabilities of the contemporary media, can further exacerbate the challenges of media coverage of emergency incidents from the perspective of a fire department or other emergency service organizations. The ability of news helicopters to travel quickly to an emergency scene and capture real-time footage that can lead to inaccurate, uninformed, and misleading coverage can present an ever-present challenge to the news media and to the fire and emergency service organizations whose incidents they cover. This typically happens when a reporter is not on the ground at the incident scene or lacks knowledge regarding emergency incident scene operations. Representative examples of such irresponsible reporting were provided in chapter 1.

A similar situation can result when the public use camera phones and other video recording devices to quickly capture still and video digital images at an incident scene and send them electronically to a television station where they show the images and comment on them often without having the necessary context to ensure responsible reporting. The ability of those engaged in the use of social media to "report" on emergency incidents to frame or tell the story from their own perspective or vantage point must be recognized and, as appropriate, addressed in assuring the accuracy of the information being disseminated relative to the incident and the fire department. An effective approach that a fire department can use to anticipate, detect, and address such inaccurate reporting is to assign an individual at the incident scene to monitor both social and traditional media coverage of the incident in the interest of revealing reporting inaccuracies to the attention of the fire department so as to be addressed by the incident commander or the PIO. The importance of addressing inaccurate information or rumors in a timely manner from the incident scene cannot be overstated.

While the focus thus far in the discussion of social media has been on the issues and challenges that they present to the fire department, social media also present a significant opportunity for the fire department to package its own informational messages and instantaneously communicate them to the public and other stakeholders. The power of social media and their potential to reach a rapidly expanding audience provides fire and emergency service organizations with a remarkable communication tool in terms of effectiveness and efficiency. Social media's timeliness and low cost as a communication medium are unparalleled. It is, however, imperative to realize that the audience that can be reached through social media is limited, a pertinent example being the current usage level by the elderly.

An obvious expectation of the fire and emergency services is that they, as organizations, and their personnel will be viewed and portrayed in a fair and positive manner. Unfortunately, this has not always been the case in news coverage of emergency incidents and the actions or inactions of fire and emergency service personnel operating on the incident scene. There will unfortunately be those rare occasions where fire department operations were flawed, such as the case when a fire department that was called through mutual aid for water supply at a structure fire and upon arriving on the incident scene firefighters discovered that their water tank for some reason was empty and that they were not prepared to complete their specified assignment. It was fortunate that there were no life safety issues involved at this incident and that the local newspaper reporter and the fire department personnel had a positive working relationship and mutual respect that had developed over the course of many years and many emergency incidents. The story and photograph that appeared on "page one above the fold" the next day, while featuring the fire, made no mention of the water supply issue. Contrast that outcome with the negative videos showing confrontations of fire service personnel with the media or the public that all too routinely appear and get significant play on the Internet.

A new reality for fire service personnel and other emergency responders operating on emergency incident scenes can be the unexpected and intrusive appearance and pressure of one or a host of cameras appearing in their faces. Such actions by the traditional news media can contribute to less than desirable working relationships between the fire service and media personnel, as well as compromising the ability of each to successfully perform given roles and responsibilities while contributing to timely, accurate, comprehensive, and professional news coverage of emergency incidents. In large part such unfortunate issues can be addressed through the provision of training and the development of relationships before the occurrence of an incident that positions the fire service and media personnel to be on the same page in terms of understanding how they can work in cooperation to ensure that the emergency situation is properly resolved and accurately reported. Many fire department PIOs have also discovered that taking time to deal with members of the public who are capturing digital images at the scene of an emergency incident may be just as important as their role in dealing with the news media from the standpoint of maintaining and enhancing the department's reputation and image.

As with other public and governmental agencies and organizations, by their very nature fire and emergency service organizations exist within a "fishbowl" in terms of visibility. Reporting that is inaccurate, misleading, or confusing can undermine their reputation and standing in the community they serve, as well as the formal and informal support that they receive from that community. A fire department thus not only has a vested interest in responsible reporting but a responsibility to work with the media to ensure this outcome of the media coverage of emergency incidents or any other activity or initiative involving the fire department.

Getting Information to the Public

There will be times when in the management of an incident the fire department or other emergency service organization will seek the assistance of the news media in the dissemination of information to an involved community. The accurate dissemination of such information can be instrumental to successful incident management, as well as contributing to the life safety and well-being of the community. An example of this type of collaboration between fire and emergency services and the news media might involve a major fire or hazardous materials incident. Were that incident to occur in or near a school, nursing home, hospital, or residential apartment building, it could further contribute to the desire for timely information on the part of the public and the need for the fire department to disseminate such relevant information as evacuations and the shelters to which those individuals who were evacuated had been taken.

In cases where a decision is made by authorities to shelter in place rather than evacuate, appropriate information should be disseminated to the public. Changes in traffic patterns and detours would likewise be important information that the news media would be capable of disseminating to the public. Sharing time-sensitive information in situations like this further illustrates the importance of developing a cooperative working relationship between the media and fire and emergency service organizations. This is further discussed in chapter 6.

Enhancing Reporting through Related Stories

There are many opportunities to enhance emergency incident reporting through the incorporation of related stories while covering a particular incident. Many of these stories will relate to what for many years has been referred to as fire prevention, wherein a fire department attempts to educate the public regarding fire hazards that they face and how they can change their behavior to reduce their vulnerability to these risks. Discussing careless smoking and the absence of operating smoke detectors while reporting on a residential fire are examples of how, by working together, the fire department and the news media can seek to prevent the recurrence of similar incidents in the future.

Enhancing media coverage of incidents to include such related issues is also appropriate when there have been incident trends within the community or the incidents are of a seasonal nature. Sidebar commentary or stories on portable heaters or carbon monoxide incidents would be appropriate examples in the winter months, whereas swimming pool safety and the dangers of fireworks would lend themselves to coverage in the summer (fig. 2–2).

Fig. 2–2. Fire department representative discussing barbecue grill safety. (Courtesy of Wilson Fire/Rescue Services.)

Many communities with older infrastructure are experiencing an increase in gas leaks. Coverage of these stories could include what to do as well as what not to do if you suspect a gas emergency. The situations mentioned here are representative of many topics that could be incorporated in the enhanced coverage of a particular emergency incident or developed as feature sidebar stories. Ideas for such stories are provided for the various types of emergency incidents considered in part 2 of this book. These coverage suggestions are designed to serve as ideas for both fire department and media personnel.

Community Risk Reduction

Many of the story ideas presented in part 2 reflect the move of progressive contemporary fire and emergency service organizations from a limited focus such as fire prevention to an all-encompassing approach to community risk reduction. An example of this is the coverage of stories that address fall prevention of the elderly, whose percentage of the U.S. population is expected to increase significantly in the next few years and to yield a corresponding increase in the demand for services, particularly in terms of emergency medical services, that fire and emergency service organizations must anticipate and prepare for. A key to success in community risk reduction will most certainly involve fire and emergency services and their media counterparts embracing and proactively addressing these issues through coordinated education and information dissemination activities.

Achieving Success Together

Through collaboration, fire departments and the news media can ensure that reporting on emergency incidents within the communities that they serve is responsible and responsive to the informational needs of community stakeholders. Understanding the challenges that each faces in enacting their responsibilities on an incident scene and their respective operations will contribute to the knowledge, empathy, and mutual respect between fire departments and the news media that will contribute to enhanced reporting that is timely, accurate, comprehensive, and professional.

The focus of this chapter is on the fire department perspective of the coverage of emergency incidents. It is designed to enable media representatives to see things from the perspective of a fire department or other emergency service organization. It likewise presents an educational agenda that fire and emergency service personnel can utilize in educating the media prior to the occurrence of an emergency situation. The next chapter considers the other side of the equation and thus provides fire and emergency service personnel with an understanding of the roles and responsibilities of the news media covering an emergency incident, along with some of the challenges that they may face in completing their assignment of getting and reporting the story.

Together, these companion chapters are designed to allow the two parties to successful and responsible reporting of emergency incidents to more fully understand each other and consequently to develop empathy with respect to each other's challenges. Through this understanding, and the insights provided in part 2 of this book, those involved in this mission-critical reporting will enhance the effectiveness of this reporting, as well as their preparedness and comfort level in covering these important incidents within a community.

Chapter Questions

1. Discuss the bias for action shared by both emergency service and media organizations.
2. Relate the process utilized in managing an emergency incident.
3. Identify, in order of importance, the priorities of incident management.
4. Discuss the focus of incident management and operations during the initial stage of an emergency incident.
5. Discuss the elements of successful emergency incident media coverage.
6. Discuss the role and importance of information dissemination to the public during an emergency incident.
7. Explain how relevant sidebar stories can enhance emergency incident media coverage.
8. Discuss the proactive approach to community risk reduction.
9. Discuss how the parties to successful media coverage can work together to ensure effective incident management and related incident media coverage.

Chapter 3

Emergency Incident Coverage—The Media Perspective

Chapter Objectives

- Relate and explain the expectations of the news media when covering an emergency incident.
- Discuss how the expectations of the media differ by media type.
- Discuss how advances in technology have changed news reporting.
- Discuss the needs of reporters in covering an assigned emergency incident.
- Discuss the possible relationship and dynamics that may exist between an emergency service organization and local news organizations.

It has been said that there is often more than one way or perspective from which to view things. That is certainly the case when it comes to the media coverage of emergency incidents. The stakeholder expectations that this coverage be not only timely, but also accurate, comprehensive, and professional, were introduced in chapter 1. The matter of emergency incident coverage was further considered from the perspective of a fire department or other emergency service organization in the second chapter. This third corresponding chapter is intended to enhance the reader's understanding of the overall context of media coverage of emergency incidents by considering things from the standpoint of the news media.

An appropriate starting point for our examination of the media perspective of emergency incident coverage is to recognize that both parties responding to an emergency incident—the fire department and the news media—have important roles with respect to the services that they provide to the community. Both are without question committed to the effective, efficient, and safe resolution of emergency situations; both must likewise commit to ensuring that the associated media coverage of emergency incidents is accurate, comprehensive, professional, and timely. The stakeholders that they together serve count on these institutions in our communities to serve as their eyes and ears as they gather, assemble, and disseminate stories related to breaking news events, such as emergency incidents, or about the fire department and other emergency service organizations on which they count to protect their lives and property.

The Essence of News Coverage

Developing an understanding of the media perspective of emergency incident coverage rightly begins with the basic concept of a story. The primary role of the news media is to keep the community informed by reporting on stories in which they are likely to have an interest. While these stories typically cover a wide range of subjects and issues that include such diverse reporting as community events, politics, sports, fashion, and entertainment, each of which is likely to appeal to a specific niche or audience, certain stories and topical areas of reporting tend to have broader appeal to newspaper readers, radio listeners, and television viewers, as well as the growing audience that follows social media to receive their news. The coverage of breaking news stories, including those on emergency incidents, falls into the category of stories that are in high demand from the perspective of the news media and those for whom they gather and disseminate the breaking news.

While the content of news stories will vary from story to story, successfully developed and delivered news will also have as its foundation a number of essential elements that are commonly referred to as the "five pillars of journalism." These five essential elements of any form of successful storytelling, including news reporting on emergency incidents, are who, what, when, where, and why. Coverage of the how or triggering events of an emergency situation or incident is also important. The very nature of the occurrence of emergency incidents demands that pertinent information with respect to each of these relevant areas be included in news reporting, regardless of the nature of a given incident or the reporting news media. The importance of news reporters being prepared to effectively gather such pertinent information and fire department representatives being prepared to efficiently provide it are discussed in the next two chapters.

The concept of "newsworthiness" plays a significant role in a news organization's determination of the stories it will cover in a given news cycle, as well as its approach to covering them in terms of the length or depth of coverage and whether coverage of an incident will involve sending news media personnel to the incident scene or covering the story remotely from a newsroom or studio. An important consideration in making these coverage decisions is the relevance of a particular subject or event to the primary audience of a particular news organization. An example of this would be limited local coverage by only the local news media of a relatively minor fire or motor vehicle accident, whereas a major fire in a large residential apartment building or known place of public assembly such as a sports arena or a major transit bus accident resulting in numerous casualties would likely be picked up and covered by the national news media. Similarly, a hazardous materials incident in a remote rural area that might take a number of days to resolve would quickly become a story of wider public and media interest if it were to result in the closure of a major interstate highway or the potential for harm from a vapor cloud traveling toward a major population area. As always, even minor incidents and events can rapidly advance to the national media stage if they involve recognized personalities or celebrities. Chapter 24 of this book is devoted to the coverage of emergency incidents of national consequence.

The News Media and Their Expectations

The success of the news coverage of an emergency incident, in terms of meeting and, ideally, exceeding stakeholder expectations for this news coverage, is dependent on the actions of both fire and emergency service personnel and their media counterparts who contribute to this coverage. Fire and emergency service personnel have information that the news media needs to deliver the accurate, comprehensive, professional, and timely media coverage that all stakeholders—including the fire department and the news media—desire. The merit and necessity of establishing a cooperative working relationship in advance of an incident and utilizing that relationship throughout the incident management and media coverage of the incident must be understood.

The importance of training both fire and emergency service personnel and media representatives in the interest of understanding each other's roles, responsibilities, and challenges introduced in the previous chapter, bears repeating. The counterpart to the unfortunate reality that most reporters have little understanding of the fire departments that they cover and related incident scene operational priorities, goals, tactics, and challenges is that the same lack of knowledge and understanding just as often exists on the part of fire and emergency service personnel with respect to the news media and their work. Working together, professionals from both of these mission-critical organizations in a community can enhance their collaborative ability to improve the accuracy, comprehensiveness, timeliness, and professionalism of the news coverage of incidents and consequently eliminate inaccurate, uninformed, or misleading coverage.

Successful reporting of emergency incidents requires that involved personnel from both parties to this news coverage—the fire department and the news media—fully understand the expectations of the stakeholders of the news coverage of emergency incidents and commit to work together to ensure that these expectations are fulfilled. Just as the fire department will expect media representatives to understand the importance of fire and emergency service organizations fulfilling their responsibilities in a timely manner, the news media deserve this same courtesy from fire and emergency service personnel. It is imperative that fire and emergency service personnel recognize that the news media have a responsibility to the community in terms of the timely and accurate communication of information about an emergency incident and that the media have a legal right to be at the incident in the interest of keeping the public informed. It should also be recognized that media representatives will routinely seek out coverage positions from which to capture incident images that are in close proximity to the unfolding incident events and operations. While the media should be afforded appropriate access to an emergency incident scene, the fire department does have the responsibility of ensuring the life safety of all personnel on an incident scene, including members of the news media. It is important that all involved personnel be knowledgeable of relevant laws with respect to media access.

The institutions and entities that we today consider the media represent an interesting historical story. News reporting in the early days of this evolution might have included the reporting by a weekly local newspaper of a fire in the town stable and blacksmith shop in which six horses perished. If you were to fast-forward a few years, you would be able to not only read about the Great Chicago Fire in newspapers across the nation, but hear about it on a growing number of radio stations springing up around the country. Further time travel in the world of news reporting would reveal the advent of television and its greatly enhanced communication capabilities in the delivery of visual images to further enhance the television viewer's understanding of what took place at an emergency incident. The unimaginable and unforgettable images of the World Trade Center towers being struck by planes and subsequently falling illustrate the instantaneous, compelling power of television coverage.

Expectations Will Differ by Media Type

News coverage of emergency incidents has been traditionally provided by newspapers, radio, and television media organizations, with events of local or regional interest being covered by local media and those of national consequence or interest being covered by national news organizations. While the basics of news reporting and coverage remain the same in terms of getting and telling the story, the associated processes and challenges escalate as the coverage advances from newspapers to radio and television, and, as a story makes the transition, from local news to national news.

The challenges facing media representatives, and coincidentally the fire and emergency service personnel contributing to news coverage, will often be driven by the deadlines of the respective news organizations. While such deadlines may be generic based on the nature of a given news media, such as the deadline for print editions of daily newspapers, others may be self-imposed based on the programming schedule of a radio or television news station. It is important that media representatives educate and inform fire service personnel involved in supplying information on emergency incidents or other fire department activities to enable them to be aware of and cognizant of the importance of providing necessary and available information in a timely manner in the interest of assisting a reporter in meeting his or her deadline. This can often be done by anticipating information requests and questions in advance of speaking with reporters.

Most communities are served by local and/or regional newspapers. Some members of a community will also seek out the news and information provided by national newspapers. In terms of emergency incident coverage, these national publications will typically cover only major incidents or those incidents occurring in major cities or metropolitan areas, while the smaller daily or weekly local and regional newspapers will concentrate on local and regional news and events, with limited coverage of national stories. Most daily newspapers today are published for distribution in the morning, and, short of actually holding the presses for a late-breaking or landmark story, will have firm deadlines the evening before. Weekly publications usually have deadlines a day or two in advance of the paper's hitting the street and thus are more inclined to recap and cover the aftermath of emergency incidents that have occurred since the last issue. Newspaper coverage of emergency incidents can consist of in-depth coverage with one or more accompanying photographs or more limited coverage provided by a brief story that is limited to the basic facts of the incident. Understanding how a reporter and his or her news organization anticipate running the story, along with the relevant deadlines, will enable fire department representatives to contribute to successful news coverage of the emergency incidents to which their department responds.

Frequently, the media coverage of emergency incidents will be provided by radio stations. Radio news coverage of emergency incidents is in real time compared to newspaper coverage of these incidents. This coverage is fast-paced, usually incorporating one or more short sound bites. Radio news coverage has the potential of quickly reaching a large audience, in that interested individuals can tune to a news radio station from a vast array of locations, including their home, car, or office. Listeners will count on their news station to have and broadcast up-to-date information at all times. Thus the challenge and importance of getting accurate information to this media in a timely manner should not be underappreciated. Radio listeners will also look to their station to pass along any important alerts or information from the public safety community, such as evacuations or road closures. While it would seem that there are not deadlines in radio broadcasting, there really are and they normally fall into hourly programming blocks, but all bets are off upon the occurrence of a major emergency situation when program plans may be set aside to allow instantaneous reporting on a story of community interest.

Based on its communication capabilities, television is an ideal media for the coverage of emergency incidents. With television a viewer can be instantaneously transported to the scene of an emergency incident that is still in progress or has recently occurred. As he or she observes the visual images of the incident, the viewer is afforded the opportunity to learn more about the context and details of the incident from a reporter on the scene or an anchor in the studio talking over the video (fig. 3–1). Often images of an incident scene may be coming from a news helicopter hovering overhead, rather than from news reporters and photojournalists on the ground.

Television coverage is even more fast-paced than that of radio. While programs are normally locked about two hours before a planned newscast, the desire to run a breaking story quite frequently changes the plans and results in the station going to live coverage of a breaking story, as did all three network stations in a major city one Friday afternoon just before rush hour after a construction backhoe cracked a large gas main in a highly populated and trafficked area with the subsequent dramatic ignition of the gas. During the early stages of this incident, as all three television stations continued to air live video, there were a number of occurrences of inaccurate and misleading reporting, resulting from the news media's lack of understanding of the nature and operations at such an incident.

While most of the emergency incidents to which a fire department responds will be of interest only to a local or regional audience, and thus to the print, radio, and television broadcast media that audience looks to for their news, there will be times when a local incident gains national attention and coverage. This increase in media attention has the potential to greatly magnify the challenges of working with the media as a fire department representative. The national news media may bring their own people and equipment or rely, at least initially, on local reporters and resources. Fire and emergency service representatives often find that they are no longer in their comfort zone in terms of working with media professionals that they know, often encountering reporters and other media personnel that they have not worked with before. Similarly, the logistics of the news vans and satellite trucks that may start arriving can be overwhelming (fig. 3–2).

Fig. 3–1. Television news coverage of a breaking news story. (Courtesy of Montgomery Country Department of Public Safety.)

Fig. 3–2. News vans and satellite trucks at a major emergency incident. (Courtesy of David Paul Brown, Montgomery County Department of Public Safety.)

A related issue is how the news cycle has changed in recent years, in large part through the advent of new technologies. The independent and network-affiliated stations that traditionally broadcast several hours of news throughout the day have expanded their news broadcasts and have been joined on the news stage by networks that broadcast the news 24 hours a day, 7 days a week. While these news organizations obviously have scheduled programming, they exist and thrive on instantaneously learning of breaking news and bringing it to their viewers—with a mission and passion for airing the story before their competition.

A growing challenge in working with the media is thus the increased number of media organizations that may have interest in carrying a story, as well as the various agendas, motivations, and deadlines that each has in covering the news. While the day that all the news trucks rolled into a community may be one that fire department personnel may talk about for years into the future, it is crucial that, as hectic and challenging as that day may have been, all can sit back and feel good about the media coverage that their incident, fire department, and community received that day. While fire and emergency service personnel do not do what they do for the glory and recognition, failure to properly prepare for and fulfill their media responsibilities at any incident, particularly one of national interest and significance, can result in the fire department being unfairly portrayed in a negative light that compromises the reputation and image that it has built over the years. Proper attention must therefore be paid to media coverage at all incidents, particularly ones that have the potential of being high visibility, in terms of either being covered by national reporters and photojournalists on the ground or picked up from a network-affiliate station.

Technology and News Reporting

While technology clearly played a role in the evolution of news media from newspapers to radio to television, it also contributed to significant advances within each of these media. In the early days a newspaper reporter would gather the news in the field and return to the newsroom to prepare the story using a typewriter in preparation for an editor's marking up the story and passing it on to the typesetter who would prepare the edition for printing. Obviously, many things have changed since those days, including electronic filing of stories by reporters from the field. Similar advances have occurred in the radio and television segments of the news business, including innovations related to how the news is gathered at an incident scene, transmitted to a studio, and broadcast to an audience eager to learn about the breaking and regular news of the day. Interestingly, there have been instances where the news media have made available their technological capabilities and equipment to assist in the management of emergency incidents, such as the use of a news helicopter for incident scene surveillance.

It is now the norm for traditional news organizations—newspapers, radio stations, and television stations—to use the Internet and other social media to supplement their news dissemination, thus allowing them to expand the scope and often timeliness of their reporting. As mentioned earlier, the explosion in the popularity, availability, and use of social media can present additional challenges to contemporary fire and emergency service organizations. We live in an age where anyone with a device capable of taking still or video digital images can capture and disseminate these images themselves or give or sell them to the news media. More about these challenges is discussed in chapter 7.

The Reporter's Needs in Covering an Incident

Just as the incident commander and other fire and emergency service personnel operating at an incident scene have a job to do, so too do the reporters. While the roles and responsibilities of a reporter in covering an emergency incident are more fully considered in chapter 5, it is appropriate that this essential role in the news coverage of an emergency incident be introduced here. The reporter covering an emergency incident, whether on- or off-site, has been assigned the responsibility of covering that incident in a manner that is fully responsive to the expectations of readers, listeners, or viewers, as well as those of the news media organization itself. Each news organization wants to have the best coverage and demands that it be accurate, complete, professional, and timely. These expectations align with those of other stakeholders, including the fire department.

When an assignment editor sends a reporter to cover an incident, he or she is sending a scarce resource that is no longer available to cover other stories. This represents a decision to invest or commit an organizational resource based on the perceived newsworthiness of the given story. The reporter certainly understands that one way or another the news organization plans to run the story and is counting on him or her to gather the necessary information to properly cover the story. The accuracy of the story may largely be dependent on the reporter's ability to gain access to the appropriate fire department personnel who can fill in the details by supplying the necessary factual information. In the previously cited example of the gas leak, had the fire department not made available a PIO to conduct a news briefing and be interviewed by reporters, a reporter might have had to draw his or her own conclusions about the situation and how the fire department was handling it or to talk to members of the public who often have opinions that can prove to be far from reality.

The greater the knowledge that a reporter has of the various types of incidents to which the fire department may respond and the corresponding operations that are appropriate at each type of incident, the less chance there will be for inaccurate, uniformed, or misleading reporting. Although astute reporters will understand that fire service personnel may be too busy performing critical tasks during the early minutes of an incident to talk to them, incident commanders must also recognize that assigning a knowledgeable representative in the form of a PIO to talk to the media as soon as possible must be high on the list of priorities.

In addition to the expectations that a reporter will have to be able to gain reasonable and safe access to the incident scene that allows him or her to capture video images of the incident and to appropriate personnel assigned to work with the media, there is another crucial expectation that should not need to be stated in that it should be blatantly obvious. Earlier the point was made that the fire department should be able to expect that reporting on incidents to which it responds will be fair, objective, and unbiased. Such an expectation on the part of a fire department stands to reason and is understood and respected by most members of the news media. The same professional courtesy should be expected and received in terms of fire and emergency service personnel not affording preferential coverage opportunities to certain news media members or organizations that they do not afford to others.

Reporters Have Bad Days Too

The emergency incidents to which fire and emergency service organizations respond on a daily basis often represent really bad days in the lives of those involved in the incident—whether the victim of an automobile accident or a residential fire. On occasion these days are also not particularly good days for a fire department or its members in terms of what they were up against that day or the resulting outcomes. Members of the fire service recognize that bad days happen and that they need to understand that as they relate to those involved—whether a coworker or the victim of an accident or fire. The same holds true with reporters. Reporters have bad days too! The long, often unexpected, hours they work coupled with the distances and conditions under which they often travel and the nature of the incidents that they are assigned to cover can lead to frustration on their part as they interact with fire and emergency service personnel in covering an emergency incident.

The Media—Friend or Foe?

This and the previous chapter of this book are intended to serve as companion chapters that examine the expectations of two essential stakeholder groups with respect to media coverage of emergency incidents: the fire department and the news media. Through fully understanding the realistic expectations of their counterparts in successful media coverage, each group can position itself to succeed. The bottom line is that both parties—the fire department and the news media—are in it together. Only by preparing together before the occurrence of an emergency incident and building upon established working relationships at the incident will both parties be capable of providing the enhanced level of incident coverage that they both desire and the community that they together serve expects and deserves.

The two chapters that follow are designed to assist in that preparation by providing an understanding of the role and responsibilities of the fire department's PIO in chapter 4 and of news reporters and other media professionals in chapter 5. The importance of developing a cooperative working relationship, as well as strategies to do so, is considered in chapter 6. Through synergistically working together, fire and emergency service personnel and their news media counterparts can and will succeed in not only meeting, but also exceeding, the expectations of their communities.

Chapter Questions

1. Relate and explain the expectations of the news media when covering an emergency incident.
2. Discuss how the expectations of the media differ by media type.
3. Discuss how advances in technology have changed news reporting.
4. Discuss the needs of reporters in covering an assigned emergency incident.
5. Discuss the possible relationship and dynamics that may exist between an emergency service organization and local news organizations.

Chapter 4

The Fire Department Public Information Officer (PIO)

Chapter Objectives

- Discuss the benefits of adopting a proactive public information approach.
- Identify the roles and responsibilities of the public information officer (PIO).
- Discuss the role of the PIO at emergency incidents.
- Discuss the qualifications of a successful PIO.
- Identify professional development opportunities that are available for the PIO.

When you think about it, the successful news coverage of emergency incidents all comes down to people. We previously considered these people based on their affiliation with one or more stakeholder groups. These groups could be further classified as producers and consumers of information. The newspaper readers, radio listeners, and television viewers who seek information regarding the communities and larger world in which they live represent information consumers who desire and seek accurate, comprehensive, and timely information about various events, including emergency incidents.

Representatives of fire and emergency service organizations who gather and provide such desired information, either through the news media or directly to those who seek the information, are considered producers of information. As a consequence of their role as middlemen or brokers of information, the news media often have a dual role in that they receive information from fire and emergency service organizations and disseminate it to the consumers of that information. The focus of this and the next chapter are on the specific roles and responsibilities of those within fire departments and the news media who contribute to the accurate, comprehensive, professional, and timely news coverage of emergency incidents—namely the fire department PIO and the news reporter.

In addition to successful reporting on emergency incidents being all about people, it is likewise all about the sharing of information. A logical starting point is therefore to distinguish between *data* and *information*. Data represent raw facts about an activity, event, or situation. When data has been processed into a useful form, it is referred to as information. An illustration of this distinction could involve the response of a local fire department along with several mutual aid departments to a working building fire in a shopping center. All of the specific details or facts about what fire departments, fire apparatus, and personnel responded to the call, as well as what they each did, would be considered data and subsequently recorded in a fire incident report prepared using the National Fire Incident Reporting System (NFIRS). The summary information that a fire department representative acting as a PIO would prepare and share with the news media would represent information, which in this case was processed into a format that met the informational needs of the media and other community stakeholders.

Benefits of a Proactive Public Information Approach

Contemporary fire and emergency service organizations have two options in terms of how they address and respond to community expectations with respect to the provision of information about their organizations and the emergency incidents to which they respond. The organization can view its responsibilities and adopt communication strategies that are either proactive or reactive. Under a reactive stance, the fire department would recognize the need to respond to information requests from the news media, the public, or other organizational stakeholders, only when and if such requests are received, and essentially "pulled" by media representatives and others. In dramatic contrast, an organization approaching its public information responsibilities in a proactive manner would be committed to advance preparation in terms of training and relationship building, followed by an aggressive approach of "pushing" relevant information to the media upon the occurrence of an emergency incident and throughout its management.

It should be clear that contemporary fire departments that, for whatever reason, approach this important mission-critical area in a reactive manner are leaving the story line up to the media, which often can result in inaccurate and misinformed reporting that can compromise the department's image and reputation in the community that it serves. Thus, a proactive approach to public information should derive from, as well as support, the fire department's mission and will prove instrumental in positioning the organization to fully meet, if not exceed, the expectations of its stakeholders with respect to media coverage. Casting the fire department in a positive light through such a proactive approach to public information and media relations has the potential of yielding significant benefits in terms of enhancing the community's understanding of the fire department and the support that it consequently receives from the community. Through recognition of the importance of public and media relations, a contemporary fire department can position itself to pursue numerous opportunities that align with its mission of service to the community.

As important as recognizing the value of the provision of such information to the media and public may be, a fire department's actual success in this mission-critical area demands that it assign qualified personnel whose responsibility it will be to operationalize the department's proactive commitment to enhancing its media coverage of emergency incidents and any other stories about the department. The establishment of a PIO position and staffing it with a highly qualified and motivated individual serves as the foundation for success in this important aspect of providing fire and emergency services in the instantaneous media-intensive world in which contemporary fire and emergency service organizations exist and operate.

While the focus of this book is on media coverage of emergency incidents, a comprehensive approach to media relations and the provision of public information will also afford opportunities for the fire department to engage in related public relations and public education activities. An example is discussing the importance of having working smoke detectors during the reporting on a fire incident where residents were saved by the early warning provided by working smoke detectors or those who perished might have survived had the residence been properly equipped with working smoke detectors. Working with the news media to incorporate such related sidebar reporting into their coverage of emergency incidents affords a fire department an opportunity to successfully drive home its fire prevention message to a motivated audience who are attentive to the issue based on the accompanying news coverage of the emergency incident. The objective of such an approach is to prevent the occurrence of similar incidents in the future. Related coverage could also include strategies that the fire department is pursuing to enhance its readiness to better serve the community through the provision of advanced or specialized personnel training, the attainment of new levels of professional certification by members or accreditation of the fire department, or putting into service new apparatus and equipment destined to improve service levels, thus enhancing the department's ability to meet and exceed stakeholder expectations.

Roles and Responsibilities of the PIO

There are many prerequisites to the successful operation of a contemporary fire department that contribute to the accomplishment of its mission and goals, and to its ability to ensure the effectiveness, efficiency, and safety of its operations both on and off the scene of emergency incidents, and consequently meet the expectations of its stakeholders. Behind the scenes there are many managerial roles, activities, and processes that contribute to this desired level of success. The research of Henry Mintzberg (1973) provides a useful framework for understanding the various managerial roles with respect to a fire department's public information function and the supporting roles and activities of its PIO. Mintzberg identified 10 possible roles that a manager may be expected to perform, based on the defined responsibilities of his or her position.[1] He further categorized these roles into three role sets: interpersonal roles, informational roles, and decisional roles. The informational roles of monitor, disseminator, and spokesperson fall within the scope of a fire department's public information activities and thus represent the roles that are typically assigned to a PIO.

The PIO enacts the role of monitor while gathering relevant information in advance of, during, and following an emergency incident. The sharing of information within the fire department and related organizations invokes the disseminator role, whereas the PIO is performing the spokesperson role when engaged in an interview with a reporter or conducting a news briefing or press conference. All three of these informational roles must be performed in an accurate, consistent, timely, and professional manner. The process through which a PIO gathers and disseminates pertinent information to the media and the public is examined in chapter 6, along with strategies that fire department and media representatives can utilize to enhance their collaborative ability to achieve the superior level of news coverage that their respective organizations and community stakeholders expect.

The growing importance of reputation management, introduced earlier, is directly related to the role and responsibilities of the PIO. The daily activities, conduct, and professionalism of a department's PIO will have a significant bearing on the organization's image in the community. While the fire chief will typically be the figurehead of the fire department in a community, it is common that the PIO becomes the face of the department in the eyes of the media, and often the public. It is therefore imperative that departmental personnel tapped to serve in this essential and often highly visible role have the necessary qualifications and interest in serving in this position. Utilizing the right person(s) in this capacity can enhance the fire department's image in the community, whereas featuring an unqualified or unwilling fire department member in this visible role can result in devastating consequences for the department.

That prompts the question of whether a fire department should appoint a designated PIO in advance of emergency incidents and routinely utilize that individual if and when available, or simply wait and make a field appointment if and when the need for a PIO presents itself and is recognized. There are obviously many advantages to the proactive approach of designating a qualified individual ahead of time and utilizing that individual in a manner that allows him or her to not only serve the department in terms of public and media relations, but in turn to be better prepared to serve as the PIO at a given incident. The alternative approach is clearly reactive and certainly not in the best interest of ensuring the desired successful news coverage of emergency incidents. The benefit of having an experienced, respected PIO when circumstances are such that misunderstandings or issues regarding an emergency incident are in need of clarification stands to reason. Additionally, the appointment and use of a designated PIO reduces confusion on an emergency incident scene, as well as contributing to reporting accuracy, consistency, professionalism, and timeliness.

While the aforementioned informational roles of monitor, disseminator, and spokesperson provide a realistic overview and understanding of the core aspects of a PIO's work, it is important that each fire department delineate the specific responsibilities assigned to their PIO. These responsibilities should be defined in writing in the form of a job description that provides insights into the requisite qualifications that an individual would need to possess to successfully perform the job as defined by the department. *NFPA 1035: Standard for Professional Qualifications for Fire and Life Safety Educator, Public Information Officer, and Juvenile Firesetter Intervention Specialist*[2] delineates job performance requirements for the PIO, and defines a PIO as an "individual who has demonstrated the ability to conduct media interviews and prepare news releases and media advisories" (*NFPA 1035*, chapter 3). This professional qualifications standard further identifies the basic function of the PIO as "coordinating and disseminating the release of information to internal and external audiences and coordinating media events." (*NFPA 1035*, annex D). The NFPA 1035 Standard further delineates the primary responsibilities of the PIO as having overall responsibility for dissemination of information regarding the organization; conducting media interviews according to organizational policies and procedures; coordinating the timely dissemination of information to internal and external audiences; writing and disseminating news releases and media advisories regarding organizational incidents, events or issues; coordinating media events and news conferences; responding when possible to emergency and nonemergency locations and serving as PIO; and utilizing opportunities to inform or reinforce organization messages within the community (*NFPA 1035*, annex D). A sample PIO position description is provided in appendix G of this book.

Role of the PIO at Emergency Incidents

While PIOs will typically have a number of roles and responsibilities related to public information, public relations, or media relations that they perform in advance of an emergency incident, their contribution to the successful news coverage of emergency incidents will appropriately define our focus. The effective, efficient dissemination of accurate and complete information to affected and interested audiences must be recognized as an essential component of the successful management of any emergency incident. Likewise, the system used to organize resources and successfully manage an emergency incident must have an appropriate design and functionality that contributes to the effective, efficient, and safe resolution of the emergency situation, along with appropriate information dissemination and news coverage about the incident.

The National Incident Management System (NIMS), developed by the U.S. Department of Homeland Security, provides a comprehensive incident management system designed to contribute to the effective, efficient, and safe management of emergency incidents in accordance with the recognized incident management priorities of (1) life safety, (2) incident stabilization, and (3) property conservation.[3] While more specifics on this incident management tool and how it contributes to successful command, coordination, and control of an emergency incident is examined in chapter 8, with supporting resource information supplied in appendix H, our focus here is to consider the role of a PIO within the NIMS.

The position of PIO, along with the companion positions of safety officer and liaison officer, are classified as command staff positions within the NIMS. As such, they serve in a staff capacity and report directly to the incident commander. The utilization of a PIO at the scene of an emergency incident allows the incident commander to focus on his or her primary command responsibilities, rather than either neglecting the informational needs of the news media or other responsibilities of managing the incident. As a member of the incident commander's command staff, the PIO handles media relations on behalf of and under the direction of the incident commander, with the incident commander retaining the ultimate responsibility for authorizing the release of information regarding the incident. While there may frequently be occasions, particularly at smaller incidents, where an incident commander will be fully capable of personally handling all of the responsibilities that fall under the command function, including working with the media, the astute incident commander will realize that there will be times when that is not the case and the assistance of a PIO will be both necessary and invaluable. The necessary roles and activities of a PIO at a given emergency incident are thus based on the nature and scope of that incident, with the size of an incident, as discussed earlier, often being a determinant of the public and media interest in and perceived newsworthiness of a particular incident. Whereas on a typical day at a routine incident, the PIO would likely be interacting with the normal cast of characters from the local news media, escalation of the incident to a story of regional or national interest and/or consequence would greatly magnify the size and scope of the news media response and inquiries, as well as the corresponding challenges confronting the PIO. This would also be the case when emergency operations are operating under an area command, emergency operations center, or multi-agency coordinating entity, under which the collection, assembly, and dissemination of incident information may be coordinated in a centralized manner.

Through the utilization of a PIO, the fire department can ensure that it is providing accurate, comprehensive, and timely information to both the media and the public, as well as reducing the risk of conflicting information coming from multiple sources. By serving as the central clearing point in the collection and dissemination of information related to an emergency incident, the PIO is positioned to significantly reduce, if not eliminate, inaccurate, confusing, or misleading media coverage of emergency incidents, while ensuring that all information dissemination activities that are performed are consistent with the fire department's policies and procedures.

Together, the incident commander and the PIO represent a team responsible for ensuring that the dissemination of information about an emergency incident is accurate, comprehensive, and timely, in accordance with the expectations of the stakeholders of this coverage (fig. 4–1). Functioning as a member of the command staff, the PIO is responsible for overseeing the successful collection and dissemination of an array of relevant information on the emergency incident. The chapters of part 2 of this book are intended to serve as a resource in terms of identifying the types of information that the incident commander, or the PIO as his or her designee, should be prepared to communicate with the media through interviews and/or press conferences. While there will be times that the PIO may be directed by the incident commander to handle media interviews or press conferences, oftentimes the PIO's role will be to schedule these opportunities for the news media to interact with the incident commander. It is often the case that the media will actually prefer to speak directly with the person running the incident—the incident commander.

Fig. 4–1. PIO conferring with incident commander. (Courtesy of Denis Onieal.)

The success of the resulting media coverage of any emergency incident will in large part relate to the working relationship and interaction between the involved fire service and media representatives. The expectations that the groups have for each other were introduced in chapters 2 and 3 and are further clarified in chapter 6. The success of information collection and dissemination at a given emergency incident is likewise based on the working relationship that exists between the incident commander and the PIO, as is the ultimate success of the news coverage related to that incident. A key to success is the selection and appointment of an appropriate individual as the PIO. This individual should have the qualifications, interest, and willingness to serve in this essential role within the incident management system.

The appointment of a designated PIO or an individual to serve in this important role at a particular incident should incorporate the principles of effective delegation, whereby the individual is granted the necessary authority to successfully fulfill his or her responsibilities and in turn is held accountable for the proper performance of those assigned responsibilities. It is also appropriate that both parties within the incident management structure with responsibility for information collection and dissemination—the PIO and the incident commander—will have certain expectations with respect to their shared working relationship.

A set of realistic expectations that an incident commander would have for his or her PIO might begin with the expectation that the PIO will maintain a current awareness of all pertinent information related to the emergency incident, including related external developments, and communicate that to all appropriate personnel operating at the incident. The PIO is expected to ensure that all information dissemination regarding the incident is properly coordinated and orchestrated in the interest of promoting a positive image of the fire department. It is both expected and imperative that the PIO identify and keep the incident commander informed of all developments, issues, or concerns about the incident, including those that may develop on the part of the public or elected officials. It is likewise imperative that the PIO always ensure that the incident commander is fully informed and prepared before appearing before the media or other interested parties in the form of interviews and/or news conferences. The PIO is also be expected to properly meet his or her responsibilities as delineated under the NIMS. This involves working closely with other members of the command staff, such as the liaison officer, whose responsibility it is to manage and coordinate interactions with various organizations, agencies, and other stakeholder groups, and representing the incident commander in various off-site PIO activities such as in a joint information center (JIC) or with agency information outlets.

Correspondingly, a PIO will usually have a set of expectations with respect to the incident commander's approach to and support for public information. First and foremost, the PIO has the right to expect that the incident commander will provide the necessary guidance and support for the work of the PIO. This begins with the clear articulation by the incident commander of his or her expectations in terms of the division of responsibility for public information roles and responsibilities between these two positions—PIO and incident commander. A second reasonable expectation is that the incident commander, as needed and when appropriate, will make time available for discussions with the PIO and for interviews and/or news conferences with the news media. To successfully meet the roles and responsibilities of PIO, a position incumbent needs to receive timely and complete guidance from the incident commander with respect to the content, form, and timing of information dissemination that has been authorized by the incident commander. This includes reviewing and authorizing press releases that the PIO has prepared before they are disseminated to the media and/or the public. The PIO should also be able to expect that the incident commander will be cooperative in terms of making him- or herself available to meet with media representatives, whether through interviews with individual or pool reporters or through participating in news conferences and other types of media briefings.

Qualifications of a Successful PIO

The success of a fire department's public information program and of the individual serving in this mission-critical position is in large part determined by the underpinning foundation provided by that individual's qualifications, willingness, and interest with respect to the PIO position. While qualifications in terms of knowledge and skills are certainly an instrumental determinant of such success, so too is an individual's actual interest or willingness with respect to serving in this vital position. No matter how qualified an individual might be in terms of knowledge and skills, it is ill advised to appoint someone as PIO and put him or her in front of the press if this person really does not feel comfortable serving in this role or is unwilling to do so.

In defining the minimum qualifications for the position of PIO, the NFPA 1035 standard states that a PIO must possess "a working knowledge of media characteristics, media interview techniques, methods of disseminating information, information technology, a comprehensive understanding of the organization and its role in the community, an understanding of emergency operations, an understanding of incident command and incident management systems, legalities of public information and organization policies and procedures" (*NFPA 1035*, annex D). The qualifications that should be considered in the selection and appointment of a PIO fall into the essential categories of knowledge, skills, and attitude. A comprehensive knowledge of the fire department's mission, goals, organizational structure, territory, key personnel, resources, policies, procedures, services, and operations is essential. The PIO should also fully understand the NIMS and its implementation at the various types of incidents to which the fire department may respond. The successful PIO will also have a thorough understanding of communication principles and practices, the characteristics and processes whereby the various news media gather and report the news, and all pertinent laws related to news coverage of emergency incidents and the related rights of the media.

Communication skills are integral to the successful enactment of the roles and responsibilities of a PIO. All three of the informational roles identified by Mintzberg—monitor, disseminator, and spokesperson—necessitate that the position incumbent be skilled in both gathering and disseminating information. Thus, both well-developed listening and public speaking skills are desirable. A coordinated set of oral and written communication skills will stand the PIO in good stead. Related interpersonal skills that afford the PIO the ability to relate to and interact with people successfully, including members of the fire department, the public, the media, representatives of other agencies, and elected and appointed officials, are also extremely important attributes of a successful PIO. Sound judgment and discretion, coupled with organizational, decision-making, research, and information management skills, also prove instrumental in equipping a PIO to excel.

The traditional requisite knowledge and skills of the successful PIO now have expanded to include the ability to fully understand and utilize available technology, the Internet, and various social media in the enactment of the PIO's role and responsibilities. As discussed earlier, the coverage of emergency incidents now takes place in a highly demanding and often impatient age of instantaneous 24/7 electronic media where it is not unusual for any individual equipped with a smart phone or digital camera to initiate the coverage of an emergency incident before the arrival of the fire department and/or the traditional news media. The ability to monitor social media coverage of emergency incidents, to professionally interact with those who have captured or are attempting to capture digital images of an incident, and to proactively use technology, including the Internet and social media, to package and provide coverage that, while meeting the expectations of stakeholders, also presents the fire department in a positive and appropriate light, most certainly constitute desirable qualifications of a contemporary PIO. In many fire departments, the PIO essentially assumes the role of a reporter, preparing and broadcasting incident-related information through the use of social media.

With increasing frequency, fire departments provide news media representatives with independent video footage captured by the public to verify and confirm the information regarding an incident prior to their arrival. This illustrates the importance of professionalism, integrity, and honesty in all of the dealings that the PIO has with the media, the public, and members of other stakeholder groups, including the fire department. The true success of a PIO and of his or her department, which ideally is built over many years and the coverage of many incidents, can be instantaneously compromised if any member of the fire department, particularly the PIO or the incident commander, were to initiate the dissemination of information or respond to questions and inquiries in a manner that was not totally accurate and truthful. It all comes down to the matter of credibility and integrity, in that any attempt to cover up or distort the truth, whether about how the fire department responded to a particular call or conducted operations at that call, is not only deceptive to the media and the public, but also can permanently compromise the organization's standing, reputation, and support in the community. This is particularly true in light of the technological world of today, in which the same media and tools that can enable the fire department, or any other organization, to advance its cause in the eyes of the public, also have the potential of quickly undermining and causing the collapse of the department's image, reputation, and credibility.

The specific application of the aforementioned knowledge, skills, and attitudes that contribute to the development and growth of the positive, productive working relationships between PIOs and reporters and that enable them to collaboratively enhance the quality of the media coverage of emergency incidents are further discussed in chapter 6.

Professional Development of the PIO

The importance of an organization, such as a fire department, providing ongoing opportunities for professional development to its members, in this case the person(s) serving as departmental PIO(s), is recognized within successful contemporary organizations. This is especially true with reference to individuals whose responsibilities are as highly visible and time-sensitive as those of a PIO. The necessity of maintaining currency in one's field, as well as further enhancing one's requisite knowledge and skills, is particularly important when an individual will be looked to as a source of accurate and credible information in the time of crisis, such as the occurrence of a major emergency incident. Examples that illustrate the importance of maintaining currency and relevancy in terms of knowledge would include being thoroughly familiar with laws, regulations, and/or standards with respect to fire department staffing and response times. Likewise, an understanding of the capabilities and ability to effectively and efficiently use new technologies such as social media to get the fire department's message out would illustrate the importance of continual skill development on the part of a PIO.

Fire departments that are successful in public information and media relations recognize the importance of making relevant professional development opportunities available to their PIOs and others whose responsibilities may at times involve representing the fire department before the media or the public. In addition to generic programs and courses offered by colleges and universities, a growing number of fire and emergency service training academies offer related courses tailored to preparing PIOs to more successfully enact their roles and responsibilities in such areas as preparing press releases, providing media interviews, and conducting press conferences. The National Fire

Academy (NFA) and the Emergency Management Institute (EMI), both located on the National Emergency Training Center (NETC) campus in Emmitsburg, Maryland, offer specific and integrated courses that contribute to the skill development of fire and emergency service personnel with public information responsibilities (fig. 4–2). Relevant courses offered by the National Fire Academy include Fire Service Communications, Presenting Effective Public Education Programs, and Community Education Leadership. The Emergency Management Institute offers Public Information Officer Awareness Training, Basic Public Information Officers, and Advanced Public Information Officers courses. The classroom learning experience in these courses, including practical hands-on learning components, is further enhanced by the networking opportunities to meet and work with public information professionals from around the nation. Information on available professional development opportunities can be found at www.usfa.fema.gov.

Fig. 4–2. Professional development course at the National Fire Academy (NFA). (Courtesy of USFA Media Production Center.)

Serving as PIO at an Incident

An important theme of the previous chapter, which examined the media perspective of emergency incident coverage, was that if the media determine that a story involving an emergency incident is newsworthy and commits to covering that story, regardless of the sources that avail themselves to reporters and the information to which they have access, the story will almost always run unless preempted by another breaking story. Thus it is always in the best interest of the fire department to do all within its power and ability to make sure that the news media get the correct story from a fire department representative capable of gathering and disseminating the pertinent information and who possesses the authority and necessary skills to effectively and efficiently tell the story from the informed perspective of the fire department. Experience has shown that when fire departments take ownership for collaboratively working with the media in their coverage of emergency incidents, the resulting news coverage meets stakeholder expectations in terms of being not only timely, but also accurate, comprehensive, and professional. The news media typically stands ready, willing, and able to provide quality coverage of the incidents to which fire and emergency service organizations respond.

The PIO plays an essential role in the successful coverage of emergency incidents in terms of serving as an information conduit between the fire department and news media. The significance of this role in maintaining and enhancing the reputation and image of the fire department in the community bears repeating. While a community may desire to have confidence and pride in its fire and emergency services, how a fire or emergency service organization responds to an emergency incident and enacts its mission-critical responsibilities on an incident scene is where the rubber meets the road in terms of public perception, confidence, and support.

The PIO, as a member of the command staff, is tasked with a role and responsibilities that in many ways can be just as important to the successful management of an emergency incident as the responsibilities performed by the incident commander. Achieving superior media coverage of emergency incidents, as discussed in chapter 1, should be the goal of each and every fire and emergency service organization. Through collaboration, the incident commander and PIO can ensure that they are effective and efficient in responding to the information needs of the news media, and in turn that the news media, through collaboration with the PIO, succeed in achieving the desired level and quality of emergency incident news coverage.

Just as an incident commander begins to collect information from the time of dispatch to arrival on an incident scene, so too should a PIO. This would include information provided in terms of radio dispatches and operational reports from initial units arriving on the incident scene. In similar fashion, the initial arriving fire department units and command officers conduct a size-up of the incident scene in terms of what has happened, what is currently happening, and what is likely to happen. The PIO should do likewise in the interest of fully understanding the context and associated challenges of the incident. Upon arrival on the incident scene the PIO will report to the incident commander to receive an informational briefing on the incident. Incorporated into this briefing should be the provision of guidance to the PIO in terms of what information the PIO is authorized to disseminate to the media and the approach that will be utilized in furnishing information to the media. During this initial briefing, which may be only the first of numerous face-to-face interactions during the management of the incident, the PIO gains important information and insights about the incident, incident priorities, strategic goals and tactics, required resources, and issues and challenges of managing the particular incident.

In preparing to meet with news media representatives, the PIO should not only anticipate the questions that they may ask on behalf of their readers, listeners, or viewers, but also be prepared to provide as complete an answer to each area of inquiry as is possible at the time of each given interaction with the media throughout the incident. The chapters in part 2 of this book are designed to provide useful insights to prompt both fire department and news media personnel in terms of the various types of information that should be gathered and incorporated into successful reporting on the various types of emergency incidents. It stands to reason that there will be times in the early stages of some incidents where the information that is available to be shared with the media will be general rather than specific. It is important that the PIO seek to further clarify unanswered questions or sketchy information as soon as possible. In such situations, it is the responsibility of the PIO to be honest with the media if he or she does not yet have the requested information and to do his or her best to secure it. It is also important that all parties to emergency incident coverage recognize and understand that there will be times when information that is available will not be released, an example being the holding of information on fatalities until appropriate notifications can be made.

At many incidents the PIO will find that it is beneficial to designate an area where the media can assemble and subsequently have access to the PIO and other fire and emergency service personnel made available to the press through the PIO. Selection of an appropriate area is important in the interest of contributing to appropriate access and functionality, including conducting interviews and press conferences. The selected location of the designated media area, as well as the available access that media representatives have on the incident scene, must be coordinated with the incident safety officer in the interest of ensuring the life safety of all personnel operating at an incident scene, including members of the news media.

This purpose of this chapter was to provide an understanding of the role that a PIO plays in the successful media coverage of emergency incidents. The next chapter looks at things from the standpoint of the reporter whose job it is to cover an emergency incident—to get and report the story. Chapter 6 more fully considers how fire department and media representatives can successfully work together in advance of, during, and after emergency incidents to enhance the media coverage of those incidents and the satisfaction on the part of the stakeholders of that coverage.

Chapter Questions

1. Discuss the importance and advantages of an emergency service organization adopting a proactive public information approach.
2. Relate and explain the roles and responsibilities of the PIO.
3. Discuss the specific responsibilities of the PIO at an emergency incident.
4. Discuss the necessary qualifications to succeed as a PIO.
5. List professional development opportunities available to a PIO.

Notes

1 Mintzberg, H. (1973). *The Nature of Management Work*. New York: Harper and Row.

2 National Fire Protection Association. (2010). *NFPA 1035: Standard for Professional Qualifications for Fire and Life Safety Educator, Public Information Officer, and Juvenile Firesetter Intervention Specialist*. Quincy, MA: National Fire Protection Association.

3 U.S. Department of Homeland Security. (December 2008). *National Incident Management System*. http://www.fema.gov/pdf/emergency/nims/NIMS_core.pdf.

Chapter 5

The Reporter's Job—Get the Story!

Chapter Objectives

- Discuss the role of the reporter in covering an emergency incident.
- Discuss how technology has changed news reporting.
- Identify the elements of successful news reporting.
- Discuss the role of preparation in successful news reporting.

While the PIO plays an instrumental, contributing role in ensuring that the news coverage of emergency incidents is accurate, comprehensive, professional, and timely, it is the news reporter who has the primary responsibility for ensuring that such reporting is successful in terms of meeting these stakeholder expectations. Reporters are journalists who research and subsequently report on various events, topics, or activities that are considered newsworthy in terms of public relevance and audience interest and thus receive coverage by the news media. Learning about the occurrence of emergency incidents within one's community or the larger world is of interest to most newspaper readers, radio listeners, and television viewers and consequently to the media organizations that cover these breaking news stories.

Reporters, as mentioned earlier, serve as middlemen or information conduits, in that they represent consumers as they gather relevant information about a news story and producers as they package, report, and disseminate the story. In so doing, they serve as the eyes and ears of the public. The traditional media organizations for whom news reporters work include newspapers, radio stations, and television stations. These organizations may be local or national in scope, and may be independent or affiliated with larger news organizations. While the geographic scope of a news organization and its corresponding audience will determine the stories regarding emergency incidents that it will be interested in covering, the nature of the media—newspaper, radio, or television—will have more to do with how the story is covered.

Additional Roles in Successful Media Coverage

Although the reporter assumes perhaps the most visible and recognized role in news reporting, in terms of receiving a byline in a newspaper or being recognized or credited in radio and television coverage of a story, he or she is in reality only one entity within the contemporary news organization. As with a successful play, the featured actors are supported by many other players who perform the necessary supporting roles to ensure the success of the production that is delivered to an audience with high anticipation and expectations. The cast of characters involved in the successful gathering and dissemination of the news coverage of emergency incidents and other stories of interest will have similar as well as unique attributes based on whether the media coverage is provided by newspapers, radio, or television.

Regardless of the type of media organization or its geographic reach in terms of readers, listeners, or viewers, there are four primary roles in the news coverage of emergency incidents: assigning the story, gathering the news, packaging or preparing the news, and reporting or disseminating the news. The occurrence of an emergency incident and the news media becoming aware of that incident, through receiving official notification from involved public safety agencies, tips from the public, or through monitoring dispatch and operational radio traffic on relevant emergency service radio frequencies, triggers the process whereby a news organization considers the newsworthiness of a particular story and thus whether it has an interest in covering a given emergency incident. This determination is typically made by an individual who has the necessary knowledge, skills, and experience to make such a determination on the part of his or her news organization and the audience of readers, listeners, or viewers that it serves. Depending on the organizational structure of the group and its size, decisions regarding what stories to cover and who to assign to each story are typically made by assignment editors or news directors.

While most stories covered by the news media are assigned to a single reporter, often accompanied by a photojournalist, more significant stories may be assigned to more than one reporter, in the interest of having each reporter cover the story from a particular, but complementary, perspective (fig. 5–1). Thus, while a single television reporter and photojournalist might be assigned with a remote camera truck to respond to and provide coverage from the incident scene of a working fire in a single-family residence, a number of such reporting teams would likely be sent by the same television station to cover a fire in a large multifamily residential apartment or condominium building. This ground coverage could also be supplemented by sending a news helicopter to the scene to provide aerial coverage of the fire and its impact on traffic on adjacent major highways or within a transportation corridor.

Fig. 5–1. Field reporter and photojournalist. (Courtesy of TrafficDan Miller.)

This field coverage could further be enhanced through having fire department personnel or other recognized authorities provide related commentary on the breaking news story by talking over real-time live video of the evolving emergency incident. Such was the case in the coverage of the pipeline explosion and fire that occurred on March 23, 1994, when a 36-inch diameter pipeline broke, resulting in an explosion at the Durham Woods apartment complex adjacent to Interstate 287 in Edison, New Jersey. In addition to destroying or inflicting significant damage to 14 apartment buildings, the fire displaced more than 100 residents. The fire that on August 13, 2008, originated and quickly consumed an unoccupied residential building at Riverwalk at Millennium, an upscale apartment complex located adjacent to the Schuylkill River and several interstate highways in Conshohocken, Pennsylvania, resulting in the destruction of three buildings, damage to five others, and displacing 375 residents, is another example of the extensive news coverage that is both appropriate and provided at major emergency incidents (fig. 5–2).

Fig. 5–2. Conshohocken, PA, Riverwalk fire. (Courtesy of Eric Chobert.)

The news stories gathered by reporters, either from the incident scene or remotely from a newsroom, are prepared for dissemination to the readers, listeners, or viewers who comprise a news organization's audience by media professionals such as the newspaper editor who edits and assembles stories for inclusion in a newspaper's print or online reporting, or a news editor or producer charged with the responsibility of packaging the news stories into appropriate segments for radio or television broadcast or inclusion on the station's online news reporting. These news production activities are performed in accordance with the nature of the media and associated attributes, reporting formats, and time or space limitations.

Newspaper coverage thus involves making a series of coordinated decisions with respect to the various stories to include in a given issue of the paper, the page(s) on which a story will appear, the length of the story, whether illustrative photographs will be included, and whether sidebar or related stories will be run. The prominence and visibility of having a story run on "page one above the fold" illustrates the impact of such layout and production decisions. Similar decisions in radio and television coverage of emergency incidents relate to the time dedicated to the primary and related coverage of an emergency incident, including later subsequent coverage, and the timing of reporting on a particular story within the overall programming of the radio or television newscast. Breaking stories, as well as other stories that are featured first in a newscast, are best positioned to capture the interest and attention of the audience. Television broadcast professionals refer to such stories as running "at the top of the broadcast," while the similar terminology for such stories in the world of radio news is a story that runs "on the hour and the half hour." This preferential ordering of major breaking news stories was illustrated by the television news coverage that all of the major Philadelphia stations and KYW News Radio incorporated into their news reporting on the day of the Riverwalk fire in Conshohocken, Pennsylvania, or when a lightning strike ignited a tank containing about 36,000 barrels (1.5 million gallons) of xylene at the Eagle Point Refinery, in West Deptford, New Jersey, on July 11, 2007.

The coverage of breaking and other news stories on emergency incidents that is provided through radio and television news organizations typically incorporates not only reporting by field reporters, but also reporting and commentary provided by radio or television personalities responsible for writing and editing news stories before incorporating them into live news broadcasts. While the official titles of these positions vary within the broadcast industry and among news organizations, such broadcast professionals are typically referred to as news anchors, anchors, newscasters, or newsreaders. Their work, whether performed from a small, obscure radio studio or from a high-tech, spacious, and aesthetically appealing broadcast studio of an independent, affiliated, or network television station, is to compile and deliver news stories.

The Reporter's Job

While the role and responsibilities of a news reporter can be easily and succinctly summarized as "getting and reporting an assigned story," the challenges that may be involved in performing this role and associated responsibilities can at times be significant and highly frustrating from the reporter's standpoint. It is interesting to note that reporters face many of the same challenges that fire departments and their members face in the delivery of effective, efficient, and safe services to the communities that they protect. The reality is that the demand for fire department services is unpredictable in terms of how many incidents will occur in a given day, the nature of those incidents, and when and where the incidents will occur. The fire department must be prepared to respond whenever and wherever its services are needed in the time of an emergency.

The same is true with respect to reporters and their news organizations. There essentially is a derived demand for the media coverage of emergency incidents. Were an emergency incident not to occur, neither the services of the fire department nor the resulting news coverage by the media would be necessary. Thus the challenges of providing media coverage of emergency incidents in many ways correspond with the challenges that fire departments face in responding to and resolving these emergency incidents. Just as members of the fire department may be pressed into service for long durations, at times when they least expect it, and often under less than desirable work conditions, such as inclement weather, so too may their counterparts in the news media—particularly the field reporters who travel to an incident scene in search of the necessary information to provide accurate and informed coverage of the incident. It is thus not unusual for reporters to find themselves in high-stress situations that often involve the potential of personal harm, particularly as they seek to secure in-depth and close-up coverage of the unfolding developments of an emergency situation in the early stages of incident management during which the situation has not yet been controlled or stabilized. While there will be times that reporters will deliver the news regarding an emergency incident from the comforts of a newsroom or broadcast studio, much of the reporting that they do will originate live from the scene of an emergency incident.

Reporters, particularly those who work for larger media organizations, are often categorized as general assignment, special assignment, or reporters who work a regular beat. When the size of a news organization's staff and expertise permits utilizing the talents of reporters in this manner, this can be advantageous in terms of enhancing the quality of news coverage that the organization provides on behalf of its audience. Under such an arrangement, reporters on general assignment would cover a wide range of stories as they present themselves and are assigned by an editor. Special assignments would afford reporters the opportunity to work on feature stories of interest to the community. Reporters assigned to a particular beat, such as fire and emergency incident reporting, would, as discussed earlier, be afforded the opportunity to develop an expert understanding of the organizations and types of stories they cover, as well as to develop sources, such as the fire department's PIO, to which they can reach out for the necessary information to enhance the accuracy, comprehensiveness, professionalism, and timeliness of reporting on emergency incidents.

The successful gathering and reporting of pertinent information on an emergency has many similarities to good news reporting in general, but there are distinct differences that derive from the nature and timing of the occurrence of an emergency incident, as well as from its potential consequences to the community. We consider two news stories to illustrate this point, the first involving the news coverage of a major league sporting team and its support of a community project by having players volunteer their time to build a playground in an inner-city neighborhood and the second involving a fire occurring in the elementary school adjacent to that playground. In the first instance, the media coverage can be prearranged, covered in a manner that contributes to the development of an outstanding story, and easily delivered to the media organization well in advance of established deadlines for field reporting.

In contrast, the occurrence of the school fire triggers the dispatch of reporters and other media resources who rush to the scene in order to be the first media organization to have the story and accompanying images and to disseminate this coverage to their audience. Such unscheduled reporting may not only present the usual challenges of meeting submission deadlines, but also takes on a new dimension when the nature, scope, or significance of an emergency incident is such that the newspaper or the radio or television station reprioritizes its previously programmed broadcast segments so as to feature a high-impact breaking news story.

How Technology Has Changed News Reporting

There is no question that technology has changed our daily lives in many ways, including how the public receives news stories and information about events of interest, such as emergency incidents. Technology has changed how such information is captured and disseminated to those with an interest in learning about the occurrence of an emergency incident, its impact, and aftermath. The cadre of amateur reporters with the capability of capturing still and digital images from an emergency incident scene has expanded exponentially, as has the ability to capture such images in a timely manner, often before the arrival of either the fire department or the traditional news media.

The advent of the Internet and social media platforms and tools enable those who have captured vivid and dramatic visual images early on and throughout an incident to instantaneously disseminate these images to a limited number of family members or friends, or to a seemingly unlimited virtual audience. The impact of visual images in communication requires no explanation. That fact that video images reflect not only the progression of an incident, but also the timing of that progression, including actions taken or not taken by the fire department, should be recognized. With increasingly frequency, television news stations encourage their viewers to capture such images, without putting themselves at personal risk, and to send their images to the station for possible inclusion in the news coverage. In addition to seeking out individuals to interview on an incident scene, many reporters now proactively seek out relevant video coverage captured by either fire department personnel or the public prior to the arrival of the reporter on the incident scene. A field reporter or news anchor typically talks over and explains such images when they are incorporated into a news broadcast.

Elements of Successful News Coverage

The elements of successful news coverage that were introduced in chapter 3—who, what, when, where, why, and how—delineate the ultimate responsibility of a reporter in providing accurate, comprehensive, professional, and timely coverage of an emergency incident. Reporting on the aforementioned lightning strike at the refinery would incorporate these essential elements as follows:

- **Who:** The incident was reported by a spokesperson for Sunoco
- **What:** Lightning struck a tank containing approximately 36,000 barrels (approximately 1.5 million gallons) of xylene (a gasoline blending component)
- **When:** Shortly after 4:30 PM on July 11, 2007
- **Where:** Sunoco's Eagle Point Refinery in West Deptford, New Jersey
- **Why:** Lightning caused an ignition source
- **How:** The ignition source resulted in the occurrence of the fire involving the contents of the storage tank

A reporter has the responsibility of gathering all of the necessary information to provide accurate and complete information on this or any emergency incident in a timely manner. In addition to accuracy and comprehensiveness, newspaper readers, radio listeners, and television viewers have the right to expect that the coverage provided by the reporters and others within their news organizations will be fair and objective. The information should address all questions that the audience may have, including such things as the environmental and health impacts of the incident. Community residents deserve, expect, and demand timely reporting of such information. Residents of the involved and neighboring communities were interested in such information regarding the incidents that occurred in Conshohocken, Edison, and West Deptford, discussed earlier in this chapter.

Preparation for Successful Emergency Incident Reporting

Successful coverage of emergency incidents is grounded in the same research, organization, and communication skills that serve as the foundation of successful reporting and news coverage of other types of stories. The successful reporter and others involved in bringing the news regarding an emergency incident to their audience will conduct background research and interviews in the coverage of an emergency incident. Comprehensive coverage of a fire in an apartment building, as well as covering the current emergency situation, could also incorporate reporting on prior fires in the building or documented issues in terms of compliance with established building or fire codes. Through extensive background research, a reporter can significantly enhance the coverage of an emergency incident by providing a context through which newspaper readers, radio listeners, or television viewers can more fully understand the situation. Organizational skills likewise prove instrumental in the development and packaging of a news story.

Interpersonal and communication skills are integral to successful news gathering, in terms of conducting interviews with individuals directly involved in a given emergency incident and of sources or subject matter experts that the reporter reaches out to for further story background or insights. Communication skills are obviously of utmost importance in the preparation and dissemination of the news coverage of emergency incidents. Well-developed written communication skills contribute to the successful development of a news story, regardless of whether that story will appear in print or serve as the script for a television or radio news report.

While the requisite skills of successful emergency incident reporting are similar to those of news reporting in general, a key to survival and success in emergency incident reporting is the development of an understanding of the challenges that fire and emergency service organizations face in the early stages of the management of an emergency incident and the patience to realize how fire department and media representatives can collaborate to ensure the effective, efficient, and safe resolution of an emergency situation, as well as the accompanying superior news coverage of the incident. Strategies that both parties can utilize in advance of, during, and following an emergency incident in the interest of enhancing incident media coverage are the focus of discussion in the next chapter of this book.

Successful media coverage of emergency incidents that fully meets and, where possible, exceeds the expectations of the stakeholders of this coverage, including the fire department and the news media, is also based on the understanding that reporters have of the fire department, the various types of incidents to which it will respond, and the operations that will be conducted at each type of incident. In reality, only a fairly limited number of reporters and media professionals possess the knowledge required to develop and report these stories in the manner that their nature, scope, and impact often deserve. When you think about it, most news organizations would never consider sending someone who knew little about sports to cover a sporting event, but conversely may have a practice of assigning naïve and uninformed reporters to cover breaking stories of emergency incidents. The information provided in the various chapters of part 2 of this book is designed to serve a dual purpose of enhancing the preparation of reporters to cover emergency incidents and the quality of the resulting coverage of these increasingly important news stories.

Chapter Questions

1. Discuss the role of a reporter when covering an emergency incident.
2. Discuss how advances in technology have changed news reporting of emergency incidents.
3. Relate and explain the elements of successful news reporting of emergency incidents.
4. Discuss the importance of preparation to successful news reporting of emergency incidents.

Chapter 6

Working Together in Emergency Incident Coverage—The Fire Department and the Media

Chapter Objectives

- Discuss how successful media coverage is a shared responsibility.
- Discuss the importance of developing a positive working relationship between the parties to emergency incident coverage.
- Discuss the general principles of working with the media.
- Discuss working with the media before an emergency incident.
- Discuss working with the media during an emergency incident.
- Discuss working with the media after an emergency incident.

Thus far we have considered the expectations that stakeholders have for the media coverage of emergency incidents, the perspectives of the fire department and the media with respect to such coverage, and the respective roles of the PIO and news reporters in contributing to the accurate, comprehensive, professional, and timely media coverage of emergency incidents. A recurring theme throughout these chapters is that successful media coverage of emergency incidents that fully meets and, ideally, exceeds stakeholder expectations can only be realized through a synergistic collaboration between the parties to this coverage—the fire department and the media. This chapter considers how such a positive working relationship can be developed in advance of an emergency incident and enhance the resulting media coverage of the incident.

There are those within the fire service who would suggest that the media is a foe, rather than a friend, of fire and emergency service organizations and should thus be considered an adversary and held at arm's length. While there will certainly be a limited number of instances where the relationship between a fire department and the media is less than desirable and in some cases perhaps even hostile, that is definitely not the case in the great majority of situations and ideally should never represent reality. Contributing to such unfortunate relationships is often the perceptions of fire department members regarding whether a given media reporter or organization was accurate, impartial, objective, and appropriate in the earlier reporting on the fire department or coverage of its emergency incidents. The issue of how a fire department can avoid inaccurate and misleading reporting and address bad news stories is discussed later in the chapter.

Most members of the fire service today recognize what the media can do to support the fire department's mission, as well as its reputation and standing in the community. These enlightened fire department members correctly view the media as a potential and valuable ally of the fire department. In the challenging economic times in which we live, the perceptions that the public and elected and appointed officials have with respect to their fire and emergency services organizations in large part may be based upon media coverage. This illustrates the ever-increasing role that media coverage may play in determining the destiny of contemporary fire and emergency service organizations.

Successful Media Coverage—A Shared Responsibility

Successful media coverage of an emergency incident begins with an understanding that fire department and media personnel have certain roles and responsibilities regarding the incident. The fire department was obviously called and responded to resolve the emergency situation in an effective, efficient, and safe manner; while the media is there to get and report the story in a manner that meets the stakeholder expectations that the coverage be accurate, comprehensive, professional, and timely. While the fire department must address its primary role and responsibilities with respect to incident management, it also has a responsibility and vested interest in ensuring that the media coverage of the incidents to which it responds meets the same expectations. The perspectives of both entities to successful media coverage of emergency incidents are in some ways unique, as discussed earlier in chapters 2 and 3, but a key similarity is the desire of both parties to get the story right. Through understanding each other's roles and responsibilities on an emergency incident scene, as discussed in chapters 4 and 5, role ambiguities and conflicts can be avoided.

Through working together, fire department and media professionals can collaboratively contribute to an enhanced level of media coverage of emergency incidents. This all sounds so logical and fairly easy to make happen; it obviously is not or it would be the reality in every community. As with most things, the foundation of the successful media coverage of emergency incidents begins well in advance of a particular incident in terms of developing a positive working relationship based on mutual understanding and respect between the key players—the PIO and the news reporters. This relationship, while developed in advance of an emergency incident, facilitates the interaction that takes place between the parties during and after an emergency incident. Through collaboration they can ensure the accuracy of reporting on the incident and related public education stories, as well as the communication of related time-sensitive alerts and information to the public.

Developing a Positive Working Relationship

The development of a positive working relationship between a fire department and the media that cover its emergency incidents and related stories occurs over time and incorporates interactions that occur prior to the occurrence of an emergency incident, those that take place during the duration of the incident, and those that occur following the incident. The actions of both parties to this relationship—the PIO and the news reporter—are instrumental to the initiation, development, and nurturing of the desired positive working relationship that by its very existence makes the interactions of the parties on the day of an emergency incident much more professional and pleasant. Contrast this with the alternative wherein no efforts have been made to develop such a positive working relationship, or the case where a negative relationship has developed over time that undermines the ability of both parties to work together as well as the quality of the resulting coverage.

The development of a positive working relationship contributes to a reduction in inaccurate, confusing, or misleading media coverage through enhancing a reporter's understanding of actions taken or not taken by the fire department on an incident scene. In cases where initial reporting on an emergency incident was inaccurate or misleading, and thus in need of timely correction, such a working relationship facilitates the timely resolution, most often from the incident scene, of issues regarding the earlier coverage. This relationship is also of utmost importance in those rare situations where a fire department's actions or inaction led to its portrayal in a negative and perhaps reputation-challenging light.

The necessity and importance of such a positive working relationship likewise increases with technology advances within the contemporary world in which fire departments respond to emergency incidents, the media provides related coverage, and an anxious and often impatient audience seeks timely, if not instantaneous, information. The traditional news media, comprised of newspapers, radio, and television, face new competition in the coverage of emergency incidents and other breaking news stories from individuals, groups, and organizations that now utilize the Internet and social media to quickly disseminate breaking news and information to a vast audience who eagerly await such information. Many traditional news organizations, likewise, now supplement their primary news dissemination through the use of Internet websites and social media. While the timely dissemination of breaking news is an essential aspect of the media coverage of any breaking story, the significant downside to such information dissemination can be inaccuracies resulting from the rush to report and the lack of verification that is typically incorporated into traditional news reporting.

The characteristics and attributes of the individuals involved in the coverage of emergency incidents—the PIO and the reporter—as well as their actions serve as the foundation upon which a successful working relationship and media coverage is built. Both parties require good communication and interpersonal skills. Their individual success as well as that of their respective organizations demand that they demonstrate honesty, integrity, credibility, and ethical behavior in all of their dealings, particularly those with their counterpart organization. Professionalism at all times and in all situations, regardless of how stressful a situation may be, also plays a crucial role in their working relationship, as well as in the resulting media coverage. Both parties obviously benefit from knowing and being proficient in their job and understanding and appreciating the other's job. Just as a reporter must understand the priorities of fire department personnel operating during the early stages of an emergency incident, so too must the fire department understand the importance of timeliness and deadlines from the perspective of the reporter and the news organization that he or she represents. The importance of making appropriate fire department personnel available for media interviews and press conferences as soon as possible must likewise be understood and guide on-scene interactions with the media. That is why it often makes sense for an incident commander to delegate this responsibility to a member of the fire department qualified to serve as the PIO for that incident. An additional dimension of developing and maintaining the desired working relationship is the agreement of both parties to keep the commitments that they make to each other, regardless of whether the commitments involve agreed to interviews on public education or related topics, the promise to conduct a series of news briefings at specified times throughout the duration of an emergency incident, or to get back to a reporter with follow-up information following an incident. Failure to honor commitments compromises the necessary trust that should be incorporated into this working relationship, as well as derived from it.

Working with the Media: General Principles

An appropriate starting point in considering the general principles that serve to underpin a positive working relationship between a PIO and news reporters is the recognition that the fire department member serving as PIO has a responsibility to his or her department, as well as the other stakeholders of effective media coverage, to act in a professional manner at all times in the interest of conveying a professional image of the organization that he or she represents. This is particularly important when media coverage, in addition to conveying one's words, actions, and mannerisms, also highlights one's appearance.

The importance of being accessible to reporters bears repeating. It is always important to remember and respect the fact that the time pressures and deadlines that the media operate under are just as real as the timing involved in the rapid growth of a fire that members of the fire service learn about early in their training as recruit firefighters. It is also imperative that all reporters, as well as the news organizations they represent, be afforded equal access to an incident scene and to interviews with appropriate fire service personnel.

The PIO, as well as other fire department personnel scheduled to talk to the press through interviews or press conferences (figs. 6–1 and 6–2), including the incident commander, should prepare themselves by anticipating the questions that members of the media are likely to ask and formulate appropriate answers to these questions. In so doing, fire department representatives should be prepared to communicate accurate, factual information, rather than opinions and should never respond in any manner that falls short of being totally truthful. Both the PIO and news reporters must understand that it may be inappropriate to disclose confidential information such as personal or medical information or that associated with a pending investigation, as in the case of an incendiary fire that authorities believe resulted from an act of arson.

Fig. 6–1. Reporter interviewing the PIO

Fig. 6–2. PIO conducting a press conference. (Courtesy of Philadelphia Fire Department.)

There will be times when a PIO will lack the knowledge or information necessary to properly respond to a reporter's question raised during an interview or a press conference. The only appropriate response in such instances is to admit to not having the answer but assure the reporter asking the question that the requested information will be sought out and subsequently shared with the news media. An attempt by the PIO to engage in "freelancing" in terms of making up a plausible answer is just as undesirable as the freelancing of firefighters operating on an incident scene. There will likewise be times when, rather than attempting to answer a particular media question or inquiry, the PIO should refer the reporter to an appropriate source of technical expertise or jurisdiction. Examples of this would be referring questions on a fire investigation to the fire marshal or other investigative agency, and those related to compliance with building or fire codes to the appropriate municipal agency. It is imperative that the PIO only promise additional information that he or she is confident can be delivered and to fulfill all such commitments to the media in a timely and professional manner.

It is always important to consider the audience when formulating statements or responses during interviews or press conferences, to refrain from the use of technical words and jargon that the audience would possibly not understand, and to remember that words mean different things to different people in different settings. For example, within the fire service the word "plug" could mean something different to the fire investigator investigating an electrical fire, the hazardous materials technician attempting to stop a leak from a container, or a fire apparatus driver/operator in search of a fire hydrant from which to establish a water supply.

Those representing the fire department in interactions with the media must understand that the entirety of such conversations is fair game in terms of the subsequent reporting of a story. All communications with the media should therefore be considered "on the record," and fire department representatives should guard their speech and only communicate that information that they have been authorized to release to the media and the public.

Working with the Media: Before an Emergency Incident

The need to prepare in advance of emergency incidents is well recognized within fire departments. In addition to making sure that all necessary apparatus and equipment is in a state of operational readiness, the progressive contemporary fire department also recognizes the importance of preparing its personnel so that they have the necessary knowledge and skills to enact their respective roles and responsibilities. Through conducting preplanning visits and familiarization tours, the fire department likewise develops an understanding of the occupancies within its jurisdiction and is able to preplan the strategies and tactics available to manage emergency incidents when they occur at these facilities.

The appointment of an individual(s) to serve in the role of PIO is an essential starting point in developing a working relationship with the media in advance of emergency incidents. In so doing, the department can not only identify appropriate individuals with the requisite qualifications to successfully enact the responsibilities of PIO, but also provide opportunities to designated individuals to engage in related professional development opportunities to develop their skills in such crucial areas as preparing press releases, conducting interviews with the media, or conducting press conferences. The ability to maintain one's composure in a high-stress environment, such as the scene of a major emergency, while reporters and photojournalists aggressively invade one's personal space with the tools of their trade—microphones, tape recorders, and cameras—can be developed through appropriate training and extensive practice.

Just as those responsible for the strategic and tactical management of an emergency incident benefit from preparing preplans and performing practice exercises, so too can a PIO develop a media briefing book containing factual information for use during the media coverage of an emergency incident. Take an example of a fire in a high-rise residential building for seniors. In addition to information regarding the building and its occupants, the PIO would benefit from background information, such as community demographics.

The essence of the positive working relationship that is advocated throughout this book all comes down to the involved individuals and the relationships that they have built over time that enable them to deliver the desired superior news coverage of emergency incidents. As with all relationships, this must be a two-way street in terms of being mutually beneficial to both parties and the interests of the organizations they represent and the audience to which they deliver the news coverage of these incidents. There are similarities between this relationship and the one that exists between contemporary fire departments and the organizations that they rely on for automatic or mutual aid. Just as a fire department recognizes the need to meet and train with the departments that it assists and receives assistance from in advance of an incident, so too should the parties to successful incident reporting get on the same page in terms of knowing and developing a working relationship with each other and educating each other before the occurrence of an emergency incident. The material in part 2 of this book should prove useful as the parties to this reporting relationship seek to orient and educate each other.

Working with the Media: During an Emergency Incident

While the strategic goals and tactics involved in the successful management of an emergency incident obviously vary based on the incident type, the priorities of incident management remain constant. The constants from the media perspective always involve getting the story in a timely manner and the techniques, such as interviewing fire department personnel, residents, or the public who witnessed the situation, utilized in gathering the necessary information. What varies in this case is the specific information that the media will deem pertinent based on the nature of the incident and the interest of the audience served by the media organization.

Fire department representatives, particularly the incident commander, must recognize that the media representatives who are assigned and respond to an incident scene to cover an emergency incident will "get the story" one way or another—regardless of whether the fire department makes available an appropriate, authorized spokesperson. The question really is, does the fire department want to play a proactive role in "framing" the story that is told, or is it willing to relinquish that opportunity as a consequence of not devoting the required attention to the informational needs of reporters through making available personnel that are both knowledgeable and authorized to discuss the incident with the media? The value of utilizing a designated PIO who can devote his or her full attention to this important role within overall incident management, rather than have the incident commander unsuccessfully attempt to simultaneously perform both roles, bears repeating.

An appropriate area that meets the needs of the reporters, as well as ensuring their safety, should be established in the interest of facilitating the process of keeping the media informed. Establishing the media area in a location that provides for the capturing of visual images of the incident scene is highly desirable, while ensuring that all images captured by either the news media or the public reflect proper and safe operations is imperative. A growing number of fire departments have assumed the role of being a primary source of the on-scene video coverage incorporated by local news organizations. The PIO should advise the media when and where updated information will be made available and should do everything possible to deliver updated information in accordance with the announced schedule. In so doing, the PIO becomes positioned as the recognized source of the timely, accurate, and credible information that will be sought by the media.

Working with the Media: After an Emergency Incident

While it would seem that the cooperating roles of the PIO and news reporters with respect to a particular incident would end with the termination of command by the incident commander who ran the incident, this is not always the case. There will be times that the PIO while operating on the incident scene will promise to get back to a reporter with additional information. It is also possible that post-incident developments and the resulting information that becomes available to the PIO may trigger him or her to proactively communicate that new or updated information to the news media. There will also be times that a reporter may reach out to the PIO at a later time in search of additional information regarding an incident.

While the urgency associated with resolving an emergency situation typically ends with the termination of the incident, it is important to remember that the news media seeking information for additional coverage are still working up against deadlines that their reporters must meet. All of the goodwill and credibility that a PIO and his or her fire department have accrued through the professional and proactive enactment of their responsibilities before and during an emergency incident can easily be compromised by not following up on subsequent media inquiries with the same degree of timeliness and urgency.

In summary, the development and maintenance of a positive working relationship between the fire department and the media all comes down to people. If the parties to this reporting relationship consider it as fairly unimportant and approach it in a casual or reactive way that results in their occasionally meeting as relative strangers on incident scenes, their lack of effort and preparation before an incident will likely result in a less than desirable outcome in terms of the resulting media coverage. In contrast would be the enhanced level of media coverage of emergency incidents that is possible when both parties take the time to get to know each other in advance of an incident and to understand the work and challenges that each face. The time, effort, and dedication that each commits to such an initiative will enable the fire department and the media to partner together as they pursue the excellent coverage of emergency incidents that their organizations desire and the community deserves.

Chapter Questions

1. Discuss the shared responsibility for successful media coverage that exists between emergency service and media personnel.
2. Discuss the importance of developing positive working relationships between the parties to emergency incident media coverage.
3. Identify several general principles for developing a positive working relationship between the parties to emergency incident media coverage.
4. Provide an example of emergency service personnel working with the media before an emergency incident.
5. Discuss how emergency service and media personnel can work together during an emergency incident.
6. Provide an example of emergency service personnel working with the media following an emergency incident.

Chapter 7

Emergency Incident Coverage in an Age of the Internet and Social Media

Chapter Objectives

- Discuss the impact of the Internet and social media on emergency incident coverage.
- Discuss the changing paradigms in news reporting as a consequence of new technologies.
- Identify the changing players in emergency incident coverage.
- Discuss the impact of the Internet and social media in breaking news stories.

This chapter considers how the Internet and social media have changed how the news media gathers and disseminates information about emergency incidents, a topic that was introduced in the first chapter and considered in the subsequent chapters of this book. While the traditional expectations of the stakeholders of the media coverage of emergency incidents have always included timely news reporting, the commercialization of the Internet in the 1990s and its subsequent growth in both applications and popularity in recent years have made instantaneous reporting both a possibility and a reality. As of 2013, nearly a third of the world's population has the Internet access that affords them the opportunity to greatly expand their horizons in terms of information about current events in their communities and the larger world.

The advent of social media has further enhanced the ability of interested parties, including the public and the news media, to research and secure timely information on a virtually unlimited number of events, subjects, and issues that occurr in their communities or the larger world, as in the case of incidents of national or international consequence or interest. The August 2011 earthquake that originated in Virginia inflicted significant damage in Washington, DC, and was felt throughout the Mid-Atlantic region and Northeast Corridor serves to illustrate the power of the Internet and social media in the timely dissemination of information. A number of national and international security professionals attending a professional development program on the Harvard University campus learned about the earthquake through social media communications initiated by their colleagues in Washington, DC, seconds before they felt the earthquake in Cambridge, Massachusetts. While the traditional news media provided extensive coverage on the earthquake and its aftermath, there were many individuals who first learned about and followed the earthquake through social media.

The media that the public counts on for primary or secondary information on breaking news, including emergency incidents, has dramatically changed over the past decade. A growing segment of the public who previously relied on the traditional news media—newspapers, radio, and television—for information on emergency incidents now look to the new informational outlets of social media, changing how one might define "the media" these days. The occurrence of Hurricane Irene within days after the Virginia earthquake and related dissemination of information through social media illustrate the role of the Internet and social media in the coverage of emergency incidents, in this case a weather-related emergency.

The Internet and Social Media

Although the Internet has been around for decades, only in recent years did it gain its current popularity and prominence within almost every aspect of contemporary society and daily life. This network of networks has dramatically revolutionized the means through which individuals, groups, and organizations now seek information and communicate about a host of things, including the occurrence of emergency incidents. Whereas not that many years ago an individual interested in learning about the outcome of a major sporting event or the occurrence of an emergency incident would have to wait patiently until such time as that information became available in the next issue of a newspaper, on the radio, or during a scheduled television news broadcast, almost any information that one might desire, particularly about current events, is now instantaneously available as a result of such technological advances as the Internet and the many new information dissemination applications that comprise what we now refer to as social media.

The advent of the Internet served as a springboard for many of the technological advances and information resources that are taken for granted today, such as the availability of websites. The Internet provides the technical platform that enables individuals, groups, and organizations to develop informational websites dedicated to any purpose or topic. In addition to their widespread use by businesses and nonprofit organizations, websites serve as powerful information-dissemination tools that the traditional news media routinely use to supplement their primary coverage of a vast array of stories of interest, including those related to emergency incidents. A growing number of public safety organizations, including fire departments and governmental agencies, likewise communicate to an interested public through their websites. It was commonplace for many of those in the path of Hurricane Irene to gain timely information regarding the path of the storm and its aftermath in terms of rain totals, flooding, and power outages from the websites of the traditional news media, as well as those of governmental and public safety organizations. There were many instances of the effective use of the Internet, whether through websites or communication through e-mail distribution lists, including the use of reverse 9-1-1 systems to communicate with targeted geographic populations, before, during, and after the occurrence of Hurricane Irene, including the dissemination of general information and alerts regarding emergency declarations, evacuations, and the establishment of emergency shelters, as well as targeted information specifically disseminated to residents in areas subject to flooding.

Blogs represent another source of web-based information dissemination. Blogs are interactive websites that are designed to facilitate information exchange regarding a particular topic or subject. Most blogs are established and maintained by individuals, whether as personal endeavors or as representatives of a group or organization.

Blogs provide interactive commentary, allowing those who visit the blog to leave comments and to message each other, thus differentiating blogs from static websites. While many blogs incorporate only textual information and comments, digital images, videos, and audio recordings can also be incorporated into the information available on a blog.

The saying "a picture is worth a thousand words" is certainly true when it comes to the design of a website and the inclusion of relevant digital images. While related still images of an event or story of interest can greatly enhance the communicative power of a website, the inclusion of digital video footage that also incorporates a real-time audio component can exponentially increase both communication effectiveness and persuasiveness. The video-sharing website YouTube facilitates the sharing of a diverse range of user-generated video content, including videos depicting emergency incidents.

The advent of social media, coupled with its rapid growth in popularity and use, represents the next level of information dissemination in the world today. Through various social media tools and platforms, individuals, groups, and organizations can now engage in interactive dialogue made possible through transforming web-based and mobile technologies. In 2011, social networking accounted for about 20 percent of all time spent online in the United States, and there were on average 40 million tweets per day on Twitter.[2] The growing popularity and use of social media is further illustrated by the more than one billion active users of Facebook[3] and 500 million users of Twitter reported in 2012.[4]

While it is likely that new social media platforms and tools will appear on the technological scene and perhaps gain the popularity realized by Facebook and Twitter, these two social networking services serve as the foundation for most current social media communication. Facebook, a social media networking service introduced in 2004, allows users to register, create personal profiles, add other users as friends, and exchange messages. User profiles can incorporate contact information and personal information, including personal interests. Facebook enables users to communicate with individuals identified as friends and others through either public or private messages and to utilize a chat feature. Users are also afforded the opportunity to join interest groups and "like" pages. An integral feature of Facebook, which is certainly relevant to information dissemination regarding emergency incidents, is its ability to share and disseminate digital images and videos.

Twitter, introduced in 2006, is considered an online social networking and microblogging service. Users of Twitter are limited in the messages that they can send and receive to a maximum of 140 characters. The communications between Twitter users, commonly referred to as "tweets," while typically encompassing a wide range of personal topics and interests, can also relate to breaking news and information, such as the occurrence of emergency incidents. The use of "hashtags" permits Twitter users to direct particular communications to users interested in following a particular topic. The power of Twitter to quickly communicate information, as well as motivate and mobilize individuals, has been illustrated in recent political campaigns and as a tool to organize protests. The relevance of the use of Twitter to alert the public and provide essential information, such as evacuation and sheltering directions during an emergency situation like a hurricane, is worth noting.

While comprehensive coverage of the history of information dissemination technologies that most take for granted today is well beyond the scope of this book, were such a history to be examined, certain individuals and organizations would assume a stature of prominence in terms of their contributions. Steve Jobs clearly deserves a prominent place is this history for his visionary and transformational leadership of Apple and how his passion and hard work has not only changed the paradigm, but revolutionized how we as a society interact and communicate. It is rather interesting to contemplate how these same technologies that Steve Jobs harnessed and introduced to the world were skillfully used to instantaneously introduce the new iPhone 4S to an eagerly awaiting world in early October 2011 and to bring us the tragic news of his death only days later.

While several highly successful social media networking services, including Facebook and Twitter, have been discussed, this discussion is in no way intended to serve as an endorsement for any particular social media tool; they are included based on the current role that each plays in social media. In all likelihood, other services and tools will join the social media ranks in the future and will be thus deserving of consideration by those in the fire service and the news media committed to the dissemination of timely information on emergency incidents to the public. It should once again be pointed out that while these social media communication tools clearly provide an avenue for the timely dissemination of information to either a targeted or general audience, they likewise can result in the dissemination of information that is inaccurate, confusing, or misleading when those initiating the communication do not take the time or have the ability to confirm the accuracy and credibility of the information that they share.

Changing Paradigms in News Reporting

The many advances in technology have revolutionized the way that the traditional news media gathers and disseminates information related to emergency incidents. News organizations have reinvented themselves and how they cover breaking news stories, such as those involving emergency incidents. A comparison of past and present reporting serves to illustrate how news coverage has changed in an age of technology.

Not that many years ago, newspaper reporters would go to an emergency incident scene where they would conduct interviews to gather the facts of the story, often taking photographs of the incident. They would often call their editor from the incident scene to relate pertinent information, establish a headline, and reserve space for the story. They would subsequently head back to the newsroom to compose the story using a typewriter or word processer, while any photographs taken at the incident scene were developed in a photo lab. Today, the same story could be composed, along with accompanying digital images, at the incident scene using a laptop, notebook computer, or other mobile data device, and forwarded electronically to an editor for review and revision, before being incorporated into desktop publishing activities involved in the publication of a contemporary newspaper.

The reporter dispatched by a radio station to cover the breaking story about an emergency incident would likewise have traveled to the scene to conduct interviews, often using a tape recorder to capture interviews of fire service personnel as well as members of the public who were directly involved in or witnessed the incident. Once again, he or she would typically call in the interview over traditional telephone lines or rush back to the radio station. The same coverage would now incorporate the capturing of the reporter's commentary and interviews using advanced digital recording devices in a format that allows the electronic transmission of these digital files back to the station for timely inclusion in its news coverage.

Television news organizations and their viewers have without question been the greatest beneficiaries of advances in communication technologies. The earlier days of a television news crew or an independent "stringer" rushing to an emergency incident scene to capture audio and video images, only to immediately rush the video footage he or she captured on a videotape back to the television station for timely editing and inclusion in news coverage have been replaced with the capture of high-definition images at an incident scene that can be immediately transmitted back to the station through microwave and satellite technologies. When time permits, such incident coverage will typically be edited before broadcast, but there will be times when live images from the scene of a breaking story may be incorporated into a news broadcast. The next sections consider how the contemporary media gathers incident-related information and how media organizations disseminate such information to the public.

Changing Players in Emergency Incident Coverage

It has been said that technology has "changed the game" of emergency incident media coverage. While this is obviously true from the standpoint of how information related to emergency incidents is gathered and disseminated, it is also the case in terms of the changing players that take the field in emergency incident coverage. Two additional parties—the fire department and the public—have joined the traditional news media of newspapers, radio, and television in the coverage of emergency incidents. The same technological advances that have contributed to the enhanced coverage of emergency incidents by traditional media organizations have facilitated the entry of both fire departments and private citizens into the often glamorous and highly visible world of the coverage and dissemination of information on breaking news stories, including those involving emergency incidents.

Advances in digital camera optics and technology allow anyone with a digital camera or smart phone equipped with such features to instantaneously capture and transmit digital images from the scene of an emergency incident and to essentially become a "reporter" and portray or "frame" the story from his or her particular perspective, regardless of how accurate and credible that perspective might be (fig. 7–1). The fact that the traditional news media and their audience desire immediate close-up coverage of emergency incidents further complicates things, particularly in cases where individuals on the incident scene in advance of their arrival may have had the opportunity to capture video images that in many cases are in great demand by the various news organizations. While traditional media organizations for years have encouraged their readers, listeners, or viewers to call in news tips, most now also encourage members of their audience to also send in still or video images captured at emergency incidents.

Fig. 7–1. Citizen capturing digital images of an emergency incident. (Courtesy of Bob Gliem.)

While the extensive news coverage that the traditional news media devoted to the introduction of the iPhone 4S in October 2011 would have been anticipated and expected, part of this coverage serves to further illustrate the growing interest of television news organizations in having their viewers send in videos of emergency incidents and other breaking news stories. In addition to discussing the enhanced 8 megapixel camera optics and increased processing speed of the long-awaited iPhone 4S, a number of television stations emphasized how their viewers equipped with the ability to capture such high-definition, broadcast-quality video images should send their videos to the station for possible inclusion in news coverage. It should thus not come as a surprise that the number of individuals capturing still or video images at a particular emergency incident is likely to continue to increase in the future, as will the potential reach and viewing audience of the images that they capture.

Many enlightened and media-savvy fire departments have recognized the importance of contributing to the story line and coverage of emergency incidents and have empowered their PIO to proactively initiate incident coverage and "push" or disseminate information to both the news media and the public. The Internet and associated social media tools provide an effective and efficient way for the fire department to disseminate accurate information in a timely manner. Many fire departments have their own websites that provide general background information on the fire department and its operations, as well as information related to particular emergency incidents of interest to the media and the public (fig. 7–2). A growing number use social media tools, including Facebook, Twitter, YouTube, and blogs to provide accurate and timely information to interested audiences, including the media and the public. In many cases, fire departments have become frequent contributors of the video and still images included in the media coverage of emergency incidents. The skillful use of the Internet and social media enables the PIO to tell the story from the fire department's perspective and to alert the news media to breaking news.

Fig. 7–2. Fire department website. (Courtesy of Chapel Hill Fire Department.)

Impact of the Internet and Social Media in Breaking News Stories

The Internet and social media have certainly left their mark on how breaking news stories, including those related to emergency incidents, are now covered in terms of gathering and disseminating information that is relevant to an incident and of interest to an intended audience. Just as expectations and interests of readers, listeners, and viewers of traditional media outlets will vary, so too will those of the recipients of information provided by nontraditional information sources through the Internet and social media. It should once again be acknowledged that while there are now many independent entities engaged in the collection and dissemination of information through the Internet and social media, a majority of newspaper, radio, and television organizations now supplement their primary or traditional coverage through the use of websites and various social media tools and are proactive in having members of the audience "like" their pages on Facebook. While many of the basics of news reporting in terms gathering information have not changed, it is fair to say that the Internet and social media have significantly changed the news business, as well as greatly enhancing the reach of a news organization, essentially providing it with a national or global footprint.

Social media and the Internet can play an integral role in alerting the public to the occurrence of an incident. While in the past the public would frequently first learn of the occurrence of an emergency incident well into or after an incident when they picked up the daily newspaper or tuned to a radio or television newscast, today information disseminated through social media can almost instantaneously trigger public awareness of an emergency incident, as well as the desire to receive timely informational updates. The party initially disseminating the information that alerted the public to an incident's occurrence may have an inherent opportunity to retain the interest of recipients of the original communication, assuming it can continue to provide what appears to be credible information in a timely manner. This illustrates the importance of both the traditional news media and the fire department positioning themselves prior to an incident as a primary or "go-to" source of the accurate, comprehensive, professional, and timely information that stakeholders seek.

While the Internet and social media offer many opportunities for a fire department to tell its story both on and off the incident scene, they also represent robust communication tools that, in the hands of those who do not approach gathering and disseminating information with the same dedication and professionalism as do fire departments and the traditional news media that cover the incidents to which they respond, can yield information that is inaccurate, confusing, or misleading. This could occur as a result of a neighbor capturing impressive video of a two-story house fire that initially showed smoke visible from a number of windows and from the eaves, with fire only showing from a first floor window. Continuing video coverage prior to fire department units arriving and going into service might reveal that the fire progressed to the point where fire is now showing from all of the windows on the first floor, the majority of the windows on the second floor, and is starting to appear from the attic and roof. Were the neighbor to post this video on the Internet or provide it to the news media, along with commentary on how long it took for the fire department to arrive and go into service, this incident through no fault of the fire department could portray the organization and its members in a negative light that, if left unaddressed, could compromise its reputation and support in the community.

The challenges of dealing with the news media and getting the correct story out would be further complicated if the fire department experienced water supply problems as a result of operational issues with the municipality's hydrant system or the fire department's apparatus. In either case, a skilled PIO, professionally enacting his or her role and responsibilities as described in the earlier chapters, will be essential in ensuring appropriate media coverage while managing the fire department's reputation. The desirability of assigning someone to monitor incident coverage provided by both the traditional news media and those providing coverage through social media bears repeating as does the importance of dealing with related issues in a timely and professional manner. It is imperative that fire department personnel avoid engaging in unprofessional confrontations with the originators of such incident coverage. There have, unfortunately, been too many examples of such confrontations subsequently becoming available on the Internet and through social media, often in the form of YouTube videos.

The Internet and social media have dramatically changed communication patterns in contemporary society, including the sources from and means through which a growing number of people get their news about events within their local community or the virtual community that is created by the reach of these innovative and powerful communication tools. Those media organizations that have survived and prospered have realized the potential of the Internet and social media and sought to fully leverage these new communication tools to enhance their business success and the effectiveness of their media coverage of stories of interest to the audience they serve.

While steeped in many traditions that at times have impeded the change and progress necessary to better serve the community, resting on tradition and maintaining a reactive approach to public information and media relations can be a risky strategy for a contemporary fire department to pursue. Earlier, the question of whether the news media should be viewed as a friend or a foe was considered. I trust that you now more fully recognize that the news media can be a valuable ally to the fire department and can contribute to its overall success, as well as the accuracy, comprehensiveness, professionalism, and timeliness of the media coverage of its emergency incidents.

The same is true of social media. The fire department that allows itself to become intimidated and fails to avail itself of the public information and media relations opportunities that social media offers, runs the risk of not properly positioning itself in the eyes of its various stakeholder groups. A proactive approach that values and fully utilizes the Internet and social media to promote the fire department's interests and enhance the media coverage it receives is absolutely essential in the challenging times in which we live. It all comes down to the question of to whom the fire department is willing to entrust its message, the story line of emergency incidents to which it responds, and ultimately its reputation. In reality, the Internet and social media are the tools of the trade of the successful PIO. The skillful use of these tools enables him or her to ensure the level and quality of media coverage of emergency incidents that will meet and exceed stakeholder expectations.

Chapter Questions

1. Discuss how technical advances, such as the Internet and social media, have had an impact on the coverage of emergency incidents.
2. Discuss how new technologies have changed news reporting practices.
3. Relate and explain the new players in emergency incident coverage.
4. Discuss how the Internet and social media have changed how breaking news stories are communicated to the public.

Notes

1 "World Stats." 2012. *Internet World Stats*. Miniwatts Marketing Group. June 30. http://www.internetworldstats.com/stats.htm.

2 "State of the Media: The Social Media Report 2012." 2012. *Featured Insights, Global, Media + Entertainment*. Nielsen. http://blog.nielsen.com/nielsenwire/social/2012.

3 Fowler, Geoffrey A. 2012. "Facebook Tops Billion-User Mark." *The Wall Street Journal*. October 4. http://online.wsj.com/article/SB10000872396390443635404578036164027386112.html.

4 Dugan, Lauren. 2012. "Twitter to Surpass 500 Million Registered Users on Wednesday." Mediabistro.com. February 21. http://www.mediabistro.com/alltwitter/500-million-registered-users_b18842.

PART 2

The first part of this book was dedicated to the process of media coverage of emergency incidents. The expectations that stakeholders have for media coverage in terms of accuracy, comprehensiveness, professionalism, and timeliness provided a context for understanding the attributes of successful media coverage of emergency incidents. The primary parties that contribute to the effective media coverage of emergency incidents—the fire department and the news media—were discussed, as were their differing perspectives and challenges associated with media coverage of emergency incidents. The roles and responsibilities of the fire department's public information officer (PIO) and the news reporter were examined. Media coverage of emergency incidents was discussed as a shared responsibility of the fire department and the media, as were strategies that both parties to successful media coverage can utilize to enhance their working relationship and the resulting coverage of emergency incidents. The evolution of news reporting in an age of the Internet and social media was an important theme throughout part 1.

The second part of this book is designed to serve a dual purpose and to address the needs of two audiences: the fire and emergency services and the news media. In addition to providing an understanding of the processes and systems that contemporary fire and emergency service organizations utilize in the management of emergency incidents, information is also provided regarding the various types of incidents to which they respond. Sixteen incident-specific chapters are included in the interest of enhancing the knowledge and understanding of the parties to successful coverage of emergency incidents. Each of these chapters is designed to serve as a preparation tool that emergency services and media personnel can use before the occurrence of an emergency incident and as a job aid or resource for use by both parties during the media coverage of a given emergency incident.

The information provided in each of these chapters can be categorized in terms of its intended audience, with certain information being specifically intended for fire and emergency services or the media, while other information will serve the preparation and resource needs of both parties to successful media coverage of emergency incidents. The information provided in each of these chapters addresses both the management of the incident and the news media's associated coverage. Media representatives will benefit from developing an understanding of the priorities, operational strategies and tactics, challenges, safety considerations, and resource requirements of the various types of emergency incidents. Further information intended to serve as a reference for the media in terms of understanding fire and emergency service terminology, acronyms, positions, apparatus, equipment, and the National Incident Management System (NIMS) is provided in the appendices that follow part 2.

The information provided in each of these chapters related to story sources, background, development, resources, and sidebar story ideas is intended to provide valuable understanding and insights for both parties to successful media coverage of emergency incidents. In addition to providing reporters with valuable guidance as they prepare for interviews and press conferences with emergency service representatives, this information can likewise alert these representatives, particularly the public information officer, to the topics that they should be prepared to address during interviews and press conferences with the news media. Such advance preparation is similar to the process of preplanning that is commonplace in contemporary fire departments. In essence, fire and emergency service representatives and their media counterparts can enhance the success of emergency incident media coverage through addressing these suggested areas of preplanning in advance of interviews and press conferences.

This book was written to contribute to the preparation of emergency service and media representatives in advance of an emergency incident and to serve as a resource to them during an emergency incident as they collaboratively seek to enhance the accuracy, comprehensiveness, professionalism, and timeliness of emergency incident media coverage. Attaining this laudable goal, and consequently meeting the expectations of the stakeholders of this coverage, is premised on the development of the necessary cooperative working relationship between the involved fire or emergency service and media representatives advocated in part 1 of this book, along with the shared understandings of emergency incident management, operations, and media coverage addressed in the second part of the book. Together, these two components are intended to enhance the ability of both parties to increase the effectiveness and success of the media coverage of emergency incidents.

Chapter 8

Incident Management

Chapter Objectives

- Discuss the importance of incident management in contributing to the effective, efficient, and safe resolution of emergency incidents.
- Identify the priorities in incident management.
- Discuss how an incident management system provides for successful media coverage of an emergency incident.
- Identify the elements of an incident management system.
- Discuss the role of the incident commander within the National Incident Management System (NIMS).
- Discuss the role of the public information officer within NIMS.
- Discuss the incident management process.
- Discuss the process of developing an action plan.
- Identify resources that are available to enhance emergency incident media coverage.
- Identify resources that are available regarding incident management.

Upon learning of the occurrence of an emergency incident within a fire department's response territory or a news media organization's coverage area, both organizations spring into action. The fire department, upon receipt of a request for emergency assistance and subsequent dispatch by a 9-1-1 communications center, responds to the incident scene to determine the nature and specifics of the emergency and to take the necessary actions to resolve the emergency situation in an effective, efficient, and safe manner. Unlike the fire department and other emergency response organizations, the respective media organizations, upon learning that an incident has occurred, make a decision as to whether to commit resources to its coverage, including the assignment of reporters, based on the perceived newsworthiness of the incident and its alignment with the interests of the readers, listeners, or viewers that comprise the media organization's audience.

This chapter examines the process that fire departments utilize in the management of emergency incidents, beginning with the priorities of incident management. Incident management is considered in a generic manner, with more detailed information on the specifics of handling the various types of emergency incidents addressed in the 16 incident-specific chapters that follow. While most individuals serving in a fire department's officer ranks, including the PIO, will likely be very familiar with incident management, and thus with this material, the chapter does serve as a good review for fire service personnel, as well as providing them with a useful lesson plan as they seek to educate their media counterparts to more fully understand and appreciate the processes and challenges of successful incident management. The primary intended audience of the chapter, however, is the news reporter or other media professional seeking to better understand the process of incident management, so as to incorporate this understanding into his or her reporting and to more successfully work in collaboration with fire department personnel to enhance the news coverage of emergency incidents.

Incident Management Priorities

The priorities in managing any emergency incident or situation, in priority order, are (1) life safety, (2) incident stabilization, and (3) property conservation. These recognized priorities drive all of the decisions and actions within the incident management process—regardless of the nature or type of incident. Life safety will always be the top priority that will guide all operations and actions on an incident scene. This focus on life safety will extend beyond response personnel to include the news media and the public.

Prior to the occurrence of an emergency incident and the fire department's dispatch and response, fire departments will have taken many measures to ensure their preparedness and readiness to respond to and resolve the emergency situation effectively, efficiently, and safely regardless of its nature, size, or scope. While this readiness is obviously dependent on personnel training and maintaining apparatus and equipment in a state of readiness, there are several other ways that fire departments prepare in advance of emergency incidents that, in addition to enhancing the effectiveness, efficiency, and safety of their on-scene operations, may give insights through which reporters can enhance their understanding of the fire department's operations, and in some cases incorporate this understanding to likewise enhance their coverage of emergency incidents.

Fire departments develop and implement policies and procedures designed to contribute to the effectiveness, efficiency, and safety of their operations on the scene of an emergency incident. *Policies* provide broad or general guidance, thus serving as guides for decision making. A safety policy that emphasizes that safety is the organization's priority both on and off the incident scene would illustrate the role of policies within a contemporary fire department. In contrast, *procedures* are specific guides for action that delineate how things are done. A procedure designed to track the accountability of fire department personnel operating on an incident scene would be an example of a procedure designed to support the department's safety policy. Through becoming familiar with a fire department's policies and procedures, news media representatives, particularly on-scene reporters, can better understand and report on the fire department and its operations at emergency incidents.

Most contemporary fire departments recognize the importance of understanding their community in advance of the occurrence of emergency incidents and thus engage in preplanning activities in the interest of becoming knowledgeable about such things as the occupancy, or use, of a particular building, its construction, water supply and fire protection, and other relevant considerations that enable the fire department to anticipate the challenges of responding to that location for a fire or other emergency and to formulate a tentative plan of how such emergencies would likely be handled. While primarily designed to inform fire department personnel, particularly those who may serve as incident commanders, in advance of an emergency incident and to provide guidance as they develop appropriate action plans on the occurrence of a particular incident, the information collected during the preplanning process and its documentation in a written document called a preplan or pre-incident plan can also represent background information that the PIO can share with reporters for inclusion in their reporting.

As discussed earlier, the primary means through which news reporters gather information and fire department personnel, typically the PIO and/or incident commander, disseminate information to the news media include interviews, press conferences, and press releases. There will be occasions, usually at larger incidents, where authorized fire department representatives may conduct tours of an incident scene for the media in the interest of further illustrating the nature and challenges of an incident, as well as discussing the fire department's strategic goals and the tactics that were employed in the management of the incident. Obviously, ensuring the life safety of members of the news media during such tours is imperative. The major incidents that occurred in Edison, New Jersey, and Conshohocken, Pennsylvania, discussed earlier, would represent situations where conducting a walking tour with the media might be appropriate.

Working Together in Media Coverage

While the fire department and the news media have a shared responsibility for ensuring that the media coverage of emergency incidents is accurate, comprehensive, professional, and timely, their perspectives and priorities on an emergency incident scene will obviously be somewhat different, as discussed in chapters 2 and 3, as will the roles and responsibilities of their personnel, as discussed in chapters 4 and 5. The first order of business for the fire department is to resolve the emergency situation in an effective, efficient, and safe manner, in accordance with the established priorities of life safety, incident stabilization, and property conservation. The priority of the news media will be to get the story and deliver it to their readers, listeners, or viewers in a timely manner. The strategies that the parties to successful media coverage can employ as they collaborate and enhance the media coverage of emergency incidents were examined in chapter 6. Understanding the process that the fire department uses in the management of emergency incidents, regardless of the type of incident, will further enhance the working relationship between relevant fire department and media personnel, as well as provide information and insights that reporters can use in their coverage of emergency incidents.

The fact that life safety will always be the priority in the management of any emergency incident bears repeating. Life safety will be a primary consideration of every strategic and operational decision made throughout the management of an emergency incident. Actions taken to ensure the safety of emergency response personnel will include the appointment of an incident safety officer and the utilization of a personnel accountability system. The establishment of hot, warm, and cold zones and accompanying perimeters to control entry of only authorized personnel into each respective zone contributes to the life safety of emergency responders, the media, and the public. The cold zone would be established in a safe area, removed from the immediate dangers of the incident, in which a command post and a designated area for the news media can be established. While such control zones were originally conceived for use at hazardous materials incidents, they are now used to ensure the safety of the media and the public at many other types of incidents, including scene control while operating on highways—incidents that every year result in injuries and fatalities of not only firefighters, but also emergency medical services and law enforcement personnel. Additional actions that may be appropriate to ensure public safety include the evacuation or sheltering in place of the public.

Incident Management System

The process of managing an emergency incident represents a real-time problem-solving exercise, wherein an incident commander must develop an understanding of an emergency situation as well as the actions that must be taken to resolve it. For many years, fire departments have utilized incident command systems, also referred to as incident management systems, as a tool for managing emergency incidents.

The utilization of an incident command system contributes to the effective resolution of an emergency incident through the efficient use of resources in a safe manner. An incident command system serves as an essential tool for enhancing command, coordination, and control of an emergency incident. It provides an organizational structure through which personnel, facilities, apparatus, equipment, and other resources can be integrated in a manner that contributes to the organized and synergistic management of an emergency incident.

A viable incident management system must serve the needs of an emergency incident, regardless of the type, location, or agencies involved. Stated another way, the incident management system must address all risks and all hazards. It likewise must be applicable to the needs of any jurisdiction and designed for use by and incorporate all response disciplines, including fire, emergency medical services, and law enforcement personnel and their organizations. An incident management system must be capable of successful implementation at incidents involving a single jurisdiction and agency, at those involving a single jurisdiction and multiple agencies, and at those involving multiple jurisdictions and agencies.

Incident management systems incorporate many traditional management principles and practices that contribute to their successful implementation and effective, efficient, and safe incident management in accordance with the recognized priorities of incident management. The National Incident Management System (NIMS), implemented by the U.S. Department of Homeland Security following September 2001, incorporated the existing principles and components of the earlier Incident Command System (ICS) in use throughout most of the fire service, including 14 proven management characteristics.[1] The 14 management characteristics delineated by the Department of Homeland Security as being incorporated in NIMS include common terminology, modular organization, management by objectives, incident action planning, manageable span of control, incident facilities and locations, comprehensive resource management, integrated communications, establishment and transfer of command, chain of command, unified command, accountability, dispatch/deployment, and information and intelligence management. Members of the news media interested in learning more about NIMS should consult the website of the Federal Emergency Management Agency (FEMA) at www.fema.gov/emergency/nims. In addition to providing information on the National Incident

Management System, this and related Department of Homeland Security websites (www.dhs.gov) and the U.S. Fire Administration (www.usfa.fema.gov) provide extensive information that can both enhance a reporter's knowledge of emergency incidents and contribute to his or her reporting on these incidents. These governmental agencies also provide extensive online courses, many of which may be of personal or professional interest to members of the news media.

NIMS provides for the development of an appropriate organizational structure tailored to the specific management needs of a given emergency incident. The person designated with the overall responsibility for managing a particular incident is referred to as the incident commander. The great majority of incidents will involve a single command wherein an individual, such as the fire chief or a senior officer, serves as the incident commander. In the case of large and more complex incidents, such as a major building fire in a medical facility where the transfer of patients to other facilities might become necessary, several individuals, such as the senior fire and senior emergency medical services officers, might together enact the command functions—a situation referred to as a unified command.

The Incident Commander within NIMS

The overall incident management responsibilities of the incident commander as delineated under NIMS include:

- Ensuring clear authority and knowledge of agency policy
- Ensuring incident safety
- Establishing an incident command post
- Obtaining a briefing from the prior incident commander and/or assessing the situation
- Establishing immediate priorities
- Determining incident objectives and strategy(ies) to be followed
- Establishing the level of organization needs and continuously monitoring the operation and effectiveness of that organization
- Managing planning meetings as required

- Approving and implementing the incident action plan
- Coordinating the activities of command and general staff
- Approving requests for additional resources or for the release of resources
- Approving the use of participants, volunteers, and auxiliary personnel
- Ordering demobilization of the incident when appropriate
- Ensuring incident after-action reports are complete
- Authorizing the release of information to the media

The significance of the accurate, comprehensive, professional, and timely media coverage advocated throughout this book is emphasized in the above responsibilities of the incident commander. NIMS further advocates that it is desirable "to promptly and effectively interact with the media, and provide informational services for the incident, involved agencies, and the public."

The incident commander is able to build an appropriate organizational structure corresponding to the needs of the particular emergency situation, based on the modular organization available through the incident command system. Whereas at a small or simple incident the incident commander may be fully capable of successfully managing the incident him- or herself, larger and more complex incidents will likely necessitate the establishment and staffing of a more robust organizational structure. In designing and implementing this organizational structure, the incident commander will incorporate the appropriate command and/or general staff positions. Those serving in command staff positions perform staff functions in support of the incident commander. These functions include incident safety, assigned to a safety officer; public information, assigned to a public information officer (PIO); and liaison with other agencies, assigned to a liaison officer. General staff positions, including operations, planning, logistics, and finance/administration, can be assigned to section officers, such as an operations officer, based on the specific management needs of an incident. While the two relevant positions from the standpoint of the media coverage of emergency incidents are obviously the incident commander and the PIO, an overview of the responsibilities of the other command and general staff positions is provided in appendix H, with further information on each of these positions or on other aspects of NIMS being available through the earlier referenced government websites.

The Public Information Officer within NIMS

The PIO is considered a command staff position within NIMS, and thus exists to support the incident commander in the management of an emergency incident. The role and responsibilities of the PIO as articulated in NIMS are consistent with those discussed earlier. The appointment of a qualified PIO enables an incident commander to achieve the desired effective and timely interaction with the media and dissemination of information to the public.

NIMS identifies the following responsibilities for the command staff position of PIO:

- Determine, according to direction from the incident commander, any limits on information release.
- Develop accurate, accessible, and timely information for use in press/media briefings.
- Obtain the incident commander's approval of news releases.
- Conduct periodic media briefings.
- Arrange for tours and other interviews or briefings that may be required.
- Monitor and forward media information that may be useful to incident planning.
- Maintain current information, summaries, and/or displays of the incident.
- Make information about the incident available to incident personnel.
- Participate in planning meetings.

The Incident Management Process

The management of a particular emergency incident begins when a call for emergency assistance is received by the jurisdiction's 9-1-1 center. After the necessary information is ascertained, the fire department and other appropriate agencies and resources are dispatched and respond to the incident. The dispatcher will dispatch the resources that correspond with the nature of the reported incident, following established dispatch policies, procedures, and protocols. Whereas a single fire department unit and crew might be dispatched upon receipt of an automatic fire alarm in a single-family residence, the report of a working fire in that same residence would likely trigger the dispatch of multiple units from a given fire department and in some cases the dispatch of automatic aid in terms of additional fire and/or emergency medical service units from neighboring organizations or jurisdictions. In addition to the units initially dispatched to the incident, the incident commander might call for additional mutual aid based on additional calls reporting a working fire received by the dispatch center, his or her observing a column of smoke in the sky while en route to the incident, or as a result of the size-up conducted upon arrival on the incident scene.

Upon arrival on the scene of any emergency incident, the incident commander will conduct a size-up of the incident that will serve as the basis for making a series of informed decisions throughout the management of the incident (figs. 8–1 and 8–2). The incident commander will be interested in understanding what has happened thus far, what is currently happening, and what is likely to happen. This information is necessary as the incident commander engages in a problem-oriented decision-making process designed to interrupt the sequence of events in the interest of resolving the emergency situation in an effective, efficient, and safe manner. The information gleaned through this initial size-up will prove essential to the successful management of the incident. Size-up, however, does not end after conducting the initial evaluation; rather, it needs to be an ongoing activity throughout the management of an incident through which progress is tracked so that appropriate adjustments may be made in strategies and tactics. A prime example of this would be making the determination at an appropriate point during an incident involving a working structure fire to switch from an *offensive mode*, wherein firefighting personnel were deployed to enter the building in the interest of performing search and rescue and fire suppression, to a *defensive mode,* wherein firefighting personnel are ordered to exit the building and mount an exterior attack of the fire. This would be appropriate in cases where the building was starting to lose its integrity and leaving personnel inside would expose them to significant harm were the collapse of the roof, ceilings, or floors to occur. In such instances where the decision is made to withdraw personnel and switch from an offensive to a defensive mode, a transitional period is required during which all personnel are accounted for; it is never appropriate to operate in both modes simultaneously.

Fig. 8–1. Incident commander performing incident size-up

1. Conduct incident size-up
2. Review incident priorities
   a. Life safety
   b. Incident stabilization
   c. Property conservation
3. Determine operational mode
4. Develop incident action plan
   a. Strategic goals
   b. Tactical objectives
   c. Resource assignments
5. Develop appropriate organizational structure
6. Implement incident action plan
7. Monitor and evaluate progress

Fig. 8–2. Incident management process

Developing an Incident Action Plan

The decision on the operational mode to use at a structure fire illustrates the magnitude and importance of the decisions that fall within the responsibility of an incident commander and those individuals serving within his or her organizational structure. While the loss of a building's integrity would certainly influence this decision, so too would the occupancy or use of the building. While the appropriate strategy in most situations will to be institute an aggressive firefighting attack, the presence of hazardous materials in a building and the known reactions and corresponding health and environmental effects caused when such materials interact with water may result in a decision to let the fire burn, while engaging in a defensive approach of protecting nearby exposures.

The process through which an incident action plan (IAP) is developed begins with the incident commander conducting a thorough size-up of the incident. Based on the information and insights gained through the size-up, he or she formulates strategic goals and tactical objectives for the management of the incident. The strategic goals represent the desired outcomes for the successful management of the incident, while the tactical objectives delineate the specific actions that will need to be taken or things that will need to be done if that outcome is to be achieved. Resources must be assigned and tasked with performing each of the necessary tactics (fig. 8–3).

The following example of a fire in a single-family residence illustrates this process. The fire was reported to the 9-1-1 center around 12:30 a.m. Shortly after receiving the initial call from an individual who had been driving past the residence, numerous other calls, all describing what the fire service would consider a "working fire," were received by the communications center. This information was relayed to the fire chief when he signed on radio and during his response to the incident scene. Upon arrival on the scene, the fire chief established command and gave an initial report of heavy fire on the first floor of a two-story, single-family residence, with limited fire spread to the second floor of the home. He subsequently did a quick 360-degree size-up of conditions by walking around the exterior of the house. He was met by a resident of the home who informed him which family members were out of the residence and accounted for, as well as those who were not accounted for at that time.

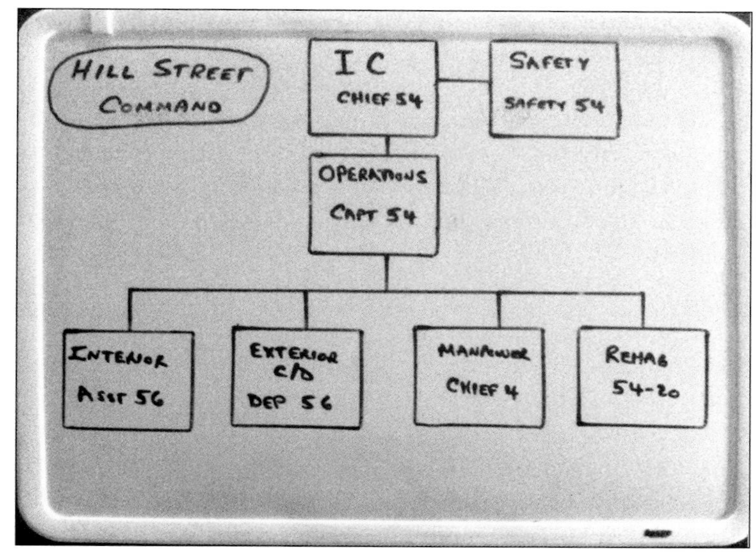

Fig. 8–3. Incident ICS organization chart

Based on this size-up, including the information from the resident, the strategic goals of rescuing the residents from the home and not incurring any injuries of firefighting personnel were identified in accordance with the recognized priorities of incident management: life safety, incident stabilization, and property conservation. The tactics required to achieve these goals were identified and included: establishing a water supply; advancing hoselines into the residence for fire suppression; conducting search and rescue for occupants; laddering the building; ventilating the building to assist in firefighting and rescue operations; controlling the utilities; and providing medical assessment, treatment, and transport, as necessary, for the residents.

The incident commander addressed the safety of personnel operating at the incident scene through the appointment of an incident safety officer (ISO) and the implementation of the department's personnel accountability system. Each of the necessary tactical objectives was assigned to appropriate responding units. One engine was assigned to establish the water supply, while the crew from a second engine handled the advancement of hoselines and conducted the search and rescue for occupants. The ladder company laddered the building with their aerial ladder and ground ladders and handled ventilation. The crew on the rescue truck assisted by controlling the utilities to the building and in search and rescue operations. Last, but certainly not least, the two emergency medical units that responded to the scene handled the assessment and treatment of the patients. In a fairly short period of time the fire department, through its organized and coordinated actions, had effectively, efficiently, and safely handled the incident and fully met the expectations of the family who lived in the residence.

While the priority, as with all incidents, had been life safety, the fire department was also successful in stabilizing the incident in a timely manner and limiting the property damage sustained by the home. The damage was significant enough, however, that the family was displaced from their residence for a period of time. Once the fire was out and the expectations of the residents had fully been met, the fire department took the time to talk to the residents about what happens next—essentially what they needed to do to return to their normal lives. In so doing, they provided the residents with a booklet available from the U.S. Fire Administration (USFA) called *After the Fire! Returning to Normal*[2] and took the time to discuss its contents with them. Thus, the fire department went from simply meeting the expectations of those residents to exceeding their expectations by demonstrating empathy for their loss and providing them a roadmap for returning to their normal lives. The information contained in this booklet can also provide valuable background information that reporters may want to incorporate in their reporting on the aftermath of a fire. This publication is available in both English and Spanish versions.

Upon returning to their fire stations, fire department personnel engaged in the necessary activities to ensure the readiness of their apparatus and equipment to respond to the next call for emergency assistance. Several days later they conducted a post-incident analysis to review the incident in the interest of learning from their experience and where possible gaining insights as to how to enhance the effectiveness, efficiency, and/or safety of their operations.

Resources for Emergency Incident Media Coverage

While this chapter provides a general overview and understanding of the process of incident management and an introduction to NIMS, more specific information on the management of the various types of emergency incidents is provided in the remaining chapters of part 2. Each of these chapters relates specific information with respect to the nature of each category of incident, as well as the strategic goals, tactics, resource requirements, and challenges typically associated with the management of each type of incident.

The appendices support these chapters. Appendices A and B provide an understanding of fire service terminology and acronyms. Appendices C, D, and E consider the resources required to successfully resolve an emergency situation in terms of personnel, apparatus, and equipment. The roles of related federal agencies and private organizations are delineated in appendix F. Appendix H provides an overview of the roles and responsibilities of the command and general staff positions within NIMS that are not considered in this and earlier chapters.

Learning More about Incident Management

While the scope of this book limits our examination of incident management, as well as NIMS, additional information is available through the earlier referenced websites of the various agencies and units within the U.S. Department of Homeland Security. A valuable source of related information for members of the fire service and the media is the Publications Department within the U.S. Fire Administration. In addition to publications related to incident management, more than 400 downloadable publications currently available at www.usfa.fema.gov/publications address such relevant topics as arson, code enforcement, community preparedness, emergency management, emergency medical services, fire prevention, fire service administration, firefighter casualties, hazardous materials, health and safety, high-risk populations, home fire safety, mass casualty incidents, multiple fire fatalities, operations and tactics, public fire education, rescue, terrorism, and wildfires. Among the available publications are technical and after-action reports on major incidents that provide important background and insights for reporters covering similar future incidents.

Likewise, PennWell/Fire Engineering publishes a comprehensive collection of fire service books and videos designed to meet the needs of contemporary fire and emergency service organizations and the professional development of their members. Many of these publications serve as valuable preparation and reference tools for news media personnel interested in expanding their knowledge and enhancing their coverage of emergency incidents. More information on these books and videos can be found at www.pennwellbooks.com/fire.html. *Fire Engineering*, a monthly professional trade publication, also proves a valuable resource for fire service and media personnel who want to stay current on events and developments in the contemporary fire service. An additional source of timely information for both the fire service and the media is the website www.fireengineering.com.

Chapter Questions

1. Discuss the role of incident management in contributing to the effective, efficient, and safe resolution of an emergency incident.
2. Relate and explain the priorities utilized in the management of an emergency incident.
3. Discuss the role of an incident management system in contributing to the successful media coverage of an emergency incident.
4. Relate and explain the elements of an incident management system.
5. Discuss the origin of the National Incident Management System (NIMS).

6. Discuss the role and responsibilities of the incident commander within NIMS.
7. Discuss the role and responsibilities of the public information officer (PIO) within NIMS.
8. Relate and explain the incident management process.
9. Discuss the importance of an action plan and its development during the incident management process.
10. List several resources that are available in the interest of enhancing the media coverage of emergency incidents.
11. List available resources regarding incident management and NIMS.

Notes

1 U.S. Department of Homeland Security. 2008. *National Incident Management System*. Emmitsburg, MD: U.S. Fire Administration. December.

2 U.S. Fire Administration. 1999. *After the Fire! Returning to Normal*. Emmitsburg, MD: U.S. Fire Administration.

Chapter 9

Accidents

Chapter Objectives

- Identify the typical categories of accidents.
- Identify the incident management priorities for accidents.
- Discuss the operational tactics utilized at accidents.
- Discuss the safety considerations associated with accidents.
- Identify the emergency service agencies that respond to accidents.
- Identify the resource requirements associated with accidents.
- Discuss the story elements associated with media coverage of accidents.

Accidents represent one of the most common types of incidents to which fire departments are dispatched. In many jurisdictions, particularly those with highly traveled major highways, the fire department will respond to just as many or more accidents each year as it does to fire-related incidents.

Typical Incidents

The types of vehicles involved in accidents, as well as the nature and extent of these accidents, varies significantly as shown in figure 9–1. The most common type of accident to which fire departments respond involves motorized vehicles designed to travel on highways or roadways, although they may not be doing so at the time of an accident, as in the case of an accident occurring in a parking lot. Passenger vehicles, which include automobiles, pickup trucks, sport utility vehicles, and buses, are the most prevalent vehicles involved in accidents (fig. 9–2). Trucks of various sizes, including tractor trailers, referred to as road freight or transport vehicles, represent another category of vehicle accident. These vehicles routinely carry a variety of shipments that often contain hazardous materials, which will influence the strategic goals and operational tactics of managing an incident. While this chapter focuses on the rescue and extrication aspects of such an incident, the management of hazardous materials incidents, including those involving fire, is considered in chapters 12, 14, and 15.

Additional transportation incidents to which a fire department may respond include rail freight and transport vehicles, including passenger and freight trains, which may or may not be transporting hazardous materials, and aircraft incidents or accidents. *Aircraft incidents* involve potential emergency situations while in flight, whereas *aircraft accidents* result from making contact with another plane, a structure, the ground, or a body of water. Given that aircraft carry passengers, cargo, and fuel, aircraft accidents will often involve the presence of hazardous materials. In general, accidents can be categorized as collisions with another vehicle, a fixed object, or a pedestrian and whether the accident involved no injuries, injuries, or fatalities. Some incidents also involve a vehicle leaving a roadway.

- Passenger vehicle accidents (automobiles, pickup trucks, sport utility vehicles, and buses)
- Rail freight or transport vehicle accidents
- Road freight or transport vehicle accidents
- Aircraft accidents or incidents
- Collision with another vehicle
- Collision with a fixed object
- Collision with a pedestrian
- Accidents without injuries or fatalities
- Accidents with injuries or fatalities

Fig. 9–1. Typical incidents: accidents

Fig. 9–2. Motor vehicle accident. (Courtesy of Robert Horton.)

Incident Priorities

As with all incidents, the priorities in the management of an accident are, in priority order, (1) life safety, (2) incident stabilization, and (3) property conservation.

Operational Overview

Review of the typical responses made by a fire department reveals the diverse array of incidents that can be classified as accidents. The nature and specifics of a given incident will thus dictate the appropriate incident management approach and actions in terms of strategic goals, tactical objectives, safety considerations, and resource requirements. The incident commander's initial size-up will provide the necessary insights to make informed decisions that will contribute to the effective, efficient, and safe management of the incident. The nature of the vehicle(s) involved, as well as their passenger(s) and/or cargo, will serve as the basis for many of these decisions. The victims of an accident may be able to self-extricate or exit easily with the assistance of emergency responders, or they may be confined within the vehicle to the point that the fire department must pop a door open using appropriate extrication tools before exit or removal is possible. In other cases, a victim may be trapped in such a manner that he or she cannot move within the vehicle or from under a vehicle and thus require fire department personnel to move or remove the part(s) of the vehicle that has him or her trapped, typically through the use of power extrication tools (fig. 9–3).

Fig. 9–3. Rescue personnel performing vehicle extrication

In cases where victims are injured, they must be assessed, treated, and transported when appropriate. While many incidents, such as a motor vehicle accident (MVA) involving a passenger car will involve a limited number of victims, multi-vehicle accidents or mass casualty incidents (MCI), such as in the case of a bus, passenger train, or aircraft accident, will, in addition to requiring the initial assessment of victims, often necessitate triaging them into categories in terms of the severity of their injuries and the urgency of their receiving treatment, given that initial resources at such an incident may not be capable of immediately handling the needs of all victims. While a fire department should always pull a safety handline when operating at an accident scene, the presence of fire or hazardous materials will dictate additional appropriate incident management actions as discussed in later chapters.

While technology has unquestionably improved many aspects of contemporary life, these same advances have contributed to challenges on an accident scene, including the presence of hazardous materials, more durable vehicle construction that can complicate vehicle extrications, dangers associated with the deployment of air bags, and hybrid vehicles that incorporate high voltages complicating firefighting, vehicle extrication and rescue, and victim removal.

Safety Considerations

The scene of an accident can present numerous life safety concerns in terms of potential harm to accident victims, emergency responders, and the public. A thorough initial and continuous size-up that incorporates scene assessment is essential in ensuring the necessary scene management to provide for the life safety of all individuals present on an accident scene. Based on this assessment, appropriate hazard control zones should be established with only essential fire department and emergency medical personnel being allowed entry into the restricted or "hot" zone. Access by the public and the media should be limited in the interest of ensuring their safety and that they do not interfere with incident operations. The control of vehicular traffic is of utmost importance to the life safety of accident victims and emergency responders and will often be accomplished through the use of large fire department vehicles blocking traffic lanes on a highway. When appropriate, such as at larger incidents, incoming resources will be staged.

In cases involving fire, the operational tactics and safety considerations regarding a hazardous materials release must also be included. These topics are further discussed in chapters 12, 14, and 15. As with all incidents, the use of a personnel accountability system and an incident safety officer (ISO) is essential. Fire department personnel involved in vehicle extrication operations are subject to various injuries, including those resulting from coming in contact with jagged metal edges of the damaged vehicle or being exposed to hazardous materials or bloodborne pathogens. The proper use of all appropriate personal protective equipment, including protection of the body, face, and eyes, will minimize the frequency and severity of such injuries and exposures. Emergency response personnel operating on a highway should always wear approved safety vests, unless they are engaged in firefighting activities where the safety vest would actually present a hazard. Properly stabilizing the involved vehicle(s) and, when appropriate, disconnecting power systems (such as the vehicle's battery), further contribute to effective, efficient, and safe operations. The protocol for handling hybrid vehicles will be different given the high voltage contained in certain vehicle systems. Potential ignition sources are always a concern when operating on an incident scene. The stretching and proper positioning of a staffed handline in the event of a problem is common practice to address this concern.

In cases where a medical helicopter is utilized, appropriate safety measures, including the establishment of a proper landing zone far enough from the incident scene, other populated areas, and electrical power lines, and under the supervision of a qualified landing zone officer, are imperative. While many accidents will be fairly minor and routine, some will not. Incidents where accident victims have suffered serious injuries and/or death, oftentimes resulting in gruesome injuries, can prove traumatic for even the most experienced emergency responders. In these cases it is important that appropriate critical incident stress debriefings and other psychological and emotional support be provided to emergency responders requiring such services. Likewise, conducting a post-incident analysis (PIA) after major or challenging incidents facilitates a learning process wherein fire and emergency service organizations and their members can learn how to enhance their future effectiveness, efficiency, and safety.

Resource Requirements

The typical response to an accident will include the fire department, emergency medical services, and law enforcement. The nature of the incident may trigger the response of governmental agencies having jurisdiction, such as the National Transportation Safety Board (NTSB) or the Federal Aviation Administration (FAA). Additional resources such as medical helicopters, towing services, and cleanup contractors may also be required.

The necessary apparatus and equipment will be based on the nature of the incident. The apparatus dispatched to typical accidents will include firefighting and rescue apparatus, and emergency medical units. Incidents involving hazardous materials may require mutual aid from hazardous materials teams, with their complement of apparatus and equipment. Likewise, aircraft accidents will trigger the dispatch of additional apparatus and units, such as airport crash trucks and foam units. The typical tools used by rescuers performing victim extrication include air bags, cribbing, cutting tools, fire extinguishers, generators, handlines, hand tools, hydraulic jacks, pneumatic tools, portable equipment, power extrication tools, power tools, prying tools, and striking tools.

Media Coverage of Accidents

Much of the local news coverage of emergency incidents will relate to the various types of accidents discussed above. Major accidents, such as bus crashes, train derailments, and airplane accidents, will often draw increased media attention and coverage and in the case of stories of national interest will present some of the challenges considered in chapter 24. In addition to authorized representatives of fire and emergency service organizations, such as the incident commander or the PIO, who will be capable of furnishing certain information about an accident, questions regarding the cause and contributing factors of an accident should be addressed by appropriate law enforcement personnel, such as a senior official or an accident investigator. It should be understood that these individuals will want to provide as much information as they can with accuracy and credibility, but will often not be able to speak to or express their opinions on the cause(s) of an accident until such time as a thorough and official investigation has been completed. In some situations where an accident reconstruction is required to properly determine the cause(s) of an accident, that information may not be available for several days. In cases where state or federal governmental agencies have jurisdiction for conducting an investigation, such as the National Transportation Safety Board (NTSB), authorized agency representatives will handle the dissemination of certain incident-related information.

While specific information on the victims of an accident will certainly be desired for inclusion in media coverage, there will be times, as discussed earlier, that the identities of victims will be withheld until such time as appropriate notifications can be made. There will also be times that such information may not be released based on the age of involved individuals. A source of specific information on the medical condition of accident victims will be authorized representatives of the medical facilities to which they were transported for treatment.

Another source of information regarding an accident will be members of the public who were involved in or witnessed an incident. In certain situations, the news media will reach out to involved organizations, such as representatives of the organization whose road or rail transport vehicle was involved in an accident. Residents of adjacent neighborhoods will often be interested in offering their thoughts on particular roadways, while official historical information on the occurrence of accidents can be secured through appropriate law enforcement agencies.

Given the impact that transportation accidents can have on highways and travel routes, those responsible for incident management will frequently request the assistance of the news media in providing traffic updates regarding road closures, congested traffic areas, and suggested detours. Information on the need to evacuate or shelter in place as a result of a transportation accident where hazardous materials are involved will also be disseminated to the public through the news media. This is another example of the cooperative media coverage discussed throughout the early chapters of this book.

The two job aids provided in figures 9–4 and 9–5 are designed to serve as resources for both fire department and media personnel. Figure 9–4 provides an overview of incident management of accidents, while figure 9–5 provides guidance with respect to media coverage of these incidents. These job aids follow a common format that is utilized throughout the incident-specific chapters in part 2 of this book. Each incident management job aid provides an overview of the incident management process for that type of incident, including typical incidents, incident priorities, operational tactics, safety considerations, responding agencies, and resource requirements, in terms of personnel, personal protective equipment (PPE), apparatus, and equipment. The nature of a particular incident will determine the appropriate operational tactics, safety considerations, responding agencies, and resource requirements. The information in the incident management job aids is designed to enhance the understanding of fire department operations on the part of media personnel, as well as serve as a tool for fire department personnel interested in assisting their media counterparts in gaining this understanding.

The media coverage job aids are likewise designed to assist reporters in getting the story of an emergency incident by providing the who, what, when, where, why, and how. In addition to serving as a useful checklist for use by reporters on an emergency incident scene, these job aids are likewise designed to assist fire department representatives as they anticipate and prepare to respond to reporter questions. Suggestions on relevant sidebar stories and reference sources are also provided. The job aids incorporated within related chapters should also be consulted as and when appropriate. Chapters 21 and 22, for example, provide important guidance in the media coverage of civilian and/or firefighter injuries and/or fatalities.

Incident Management: Accidents	
Typical Incidents	• Passenger vehicle accidents (automobiles, pickup trucks, sport utility vehicles, and buses) • Rail freight or transport vehicle accidents • Road freight or transport vehicle accidents • Aircraft accidents or incidents • Collision with another vehicle • Collision with a fixed object • Collision with a pedestrian • Accidents without injuries or fatalities • Accidents with injuries or fatalities
Incident Priorities	1. Life safety 2. Incident stabilization 3. Property conservation
Operational Tactics*	• Control zones, crowd control, extrication, fire control, landing zone, medical treatment and transport, perimeter control, personnel accountability, rescue, scene assessment, scene management, size-up, stabilization, staging, and traffic control
Safety Considerations*	• Carbon monoxide, exposure (bloodborne pathogens), extrication hazards, hazardous materials, helicopter operations and landing hazards, hybrid vehicles, ignition sources, injuries, personnel accountability, utilities, vehicle stabilization, and vehicular traffic
Responding Agencies*	• Fire department, emergency medical services (basic life support, advanced life support, and/or medical helicopter), fire police, law enforcement, and governmental agencies having jurisdiction (certain incidents)
Resource Requirements	Personnel* • Fire department, emergency medical services, fire police, and law enforcement Personal protective equipment* • Coat, pants, boots, helmet, gloves, eye protection, safety vest, and personal alert safety system (PASS) Apparatus* • Emergency medical units, firefighting apparatus, and rescue apparatus Equipment* • Air bags, cribbing, cutting tools, fire extinguishers, generators, handlines, hand tools, hydraulic jacks, pneumatic tools, portable equipment, power extrication tools, power tools, prying tools, striking tools, and traffic control equipment

Fig. 9–4. Incident management job aid. **Note:* The nature of a particular incident will determine the appropriate operational tactics, safety considerations, responding agencies, and resource requirements.

Media Coverage: Accidents	
WHO?	• Number of vehicle passengers? • Number of pedestrians? • Number of victims? • Number of and extent of civilian injuries? • Number of and cause of civilian fatalities? • Number of and extent of emergency responder injuries? • Number of and cause of emergency responder fatalities? • Personal information on injured individuals? • Personal information on fatalities? • *Have appropriate notifications been made?* • *Has information on victims been received from a credible and authorized source?* • *Has information been verified for accuracy?* • How and to which medical facilities were victims transported? • Medical condition of victims (initial and present)?
WHAT?	• Type of incident (passenger, rail freight or transport, road freight or transport, aircraft)? • Number and type of involved vehicles? • Cargo being transported? • Involved organizations (railroad, trucking company, airline)? • Damage to other vehicles? • Damage to buildings? • Damage to surroundings? • Responding agencies? • Incident priorities? • Operational tactics? • Safety considerations? • Incident management challenges? • Resource requirements (number and types of personnel, apparatus, and equipment)? • Resulting fire? • Hazardous materials involvement? • Resulting health or environmental concerns? • Evacuation or shelter in place? • Road closures or lane restrictions? • Traffic detours?
WHEN?	• Day of week? • Time of day? • Day/night?

Fig. 9–5. Media coverage job aid

WHERE?	• Incident location? • Nearby recognized landmarks (roads, buildings, etc.)? • Alternate location references (route number vs. street name)? • Proximity to past accidents?
WHY?	• Factors contributing to accident? • Human factors? • Weather-related factors? • Official determination of cause(s)? • Charges brought against responsible parties?
HOW?	• How did the accident occur? • Collision with another vehicle? • Collision with fixed object? • Collision with pedestrian? • Vehicle left the roadway?
Sidebar Story Ideas	• Airline passenger safety • Alcohol consumption by vehicle driver/operators • Distracted drivers • Elderly drivers • Established detour routes • Hybrid vehicles • Previous similar incidents • Proper child car seat installation • Proper vehicle maintenance • Seatbelt use • Stories on victims of accidents • Substance abuse by vehicle driver/operators • Weather-related accidents • Young drivers
Resources	• Centers for Disease Control and Prevention (CDC) www.cdc.gov • Federal Aviation Administration (FAA) www.faa.gov • National Highway Traffic Safety Administration (NHTSA) www.nhtsa.gov • National Transportation Safety Board (NTSB) www.ntsb.gov • United States Fire Administration (USFA) www.usfa.fema.gov

Fig. 9–5. Media coverage job aid (continued)

Chapter Questions

1. Identify the typical categories of accidents.
2. List the incident management priorities in managing accidents.
3. Discuss several operational tactics related to accidents.
4. Identify the safety concerns related to accidents.
5. Identify the emergency service and support organizations and agencies that are dispatched to accidents.
6. Discuss the resource requirements, in terms of personnel, apparatus, and equipment, associated with accidents.
7. Relate and explain the essential story elements of news coverage of accidents.

Chapter 10
Building Fires

Chapter Objectives

- Identify the typical categories of building fires.
- Identify the incident management priorities for building fires.
- Discuss the operational tactics utilized at building fires.
- Discuss the safety considerations associated with building fires.
- Identify the emergency service agencies that respond to building fires.
- Identify the resource requirements associated with building fires.
- Discuss the story elements associated with media coverage of building fires.

According to the National Fire Protection Association (NFPA) there were an estimated 484,500 structure fires in 2011. Residential structure fires resulted in 15,635 civilian injuries and 2,640 civilian deaths that year, whereas nonresidential structure fires accounted for 1,865 civilian injuries and 365 civilian deaths. The resulting direct dollar loss resulting from structure fires was $9.7 billion.[1] While a growing number of fire department dispatches are for automatic fire alarms in buildings, in the majority of such cases it will be discovered that there is no fire or need for fire department services.

Typical Incidents

Building fires, also referred to as structure fires, result from ignition of a fire and its subsequent involvement of a building and/or its contents. Building fires are typically classified by the occupancy, or present use, of a building, as shown in figure 10–1. While a number of model code systems exist, each with its own occupancy classifications, specific building and fire codes will be adopted by the authority having jurisdiction (AHJ). In addition to the number and nature of occupancy classifications varying under each classification system, the specific building uses under each occupancy class can be numerous, with only representative examples being provided herein. General occupancy classes include assembly, business, educational, industrial, institutional, mercantile, residential, storage, and miscellaneous. It is not unusual to find mixed occupancies within a building, as in the case of apartments on the floor(s) above retail stores.

- Assembly occupancies (amusement arcade, exhibit hall, house of worship, movie theatre, nightclub, restaurant, showplace, or sports arena)
- Business occupancies (offices)
- Educational occupancies (elementary or secondary school)
- Industrial occupancies (factory or processing plant)
- Institutional occupancies (ambulatory healthcare facility, boarding house, correctional facility, day-care center, detention center, healthcare facility, hospital, or nursing home)
- Mercantile occupancies (retail store, shopping mall, or strip shopping center)
- Residential occupancies (apartment building, condominium, dormitory, hotel, mobile home, motel, rooming house, row house, single-family residence, townhouse, or multi-family residence [e.g., duplex or triplex])
- Storage occupancies (bulk storage facility, distribution center, garage, storage facility, or warehouse)
- Miscellaneous occupancies (portable buildings)
- Fire location (attic, basement, chimney, high-rise, or roof)

Fig. 10–1. Typical incidents: building fires

The occupancy of a building will be a key determinant of the issues that a fire department will face in responding to a building fire. The occupancy or use of a building will determine the number of people occupying the building, the hours during which the building is occupied, and the ability of building occupants to evacuate the building without assistance. Residential buildings such as single-family homes and apartment buildings typically have a higher census during the evening than during the day and pose additional challenges when residents are asleep (fig. 10–2). Likewise, the challenges of ensuring that all residents are safely evacuated from a one- or two-family home can be significantly less than those associated with accounting for all of the residents of an apartment building, particularly a high-rise residential building. While most occupants of residential structures will be capable of evacuating themselves from a building upon being alerted to the need, others, including those with infirmities, will require assistance to do so. This is often the case in a nursing home or hospital fire, with there being times that evacuation from a hospital may be problematic based on a patient's condition or a procedure that the patient is undergoing at the time of a fire. Some hospital fire plans may thus incorporate strategies that shelter such patients in place in compartmentalized areas of the building that are unaffected by a minor fire in a remote location. A fire in an elementary school would likewise pose challenges in ensuring that all of the students and staff are accounted for after an evacuation of the building.

Fig. 10–2. Residential building fire

The focus of the above discussion has rightfully been on ensuring the life safety of building occupants in the event of a fire. Ensuring an orderly building evacuation that results in moving all involved individuals from an area where they are in harm's way to a safe and secure area will always be a priority in fire department operations. This will likewise be the case in the various other occupancies, whether an office building, a retail store, a movie theatre, a house of worship, or other occupancy. Fire department personnel will search the building to ensure that all occupants have been evacuated or relocated to areas of safe haven, when such a sheltering in place strategy is determined to be appropriate in a compartmentalized building.

The occupancy of a building may also present certain challenges with respect to the presence or use of hazardous materials that could be problematic in the event of a fire. The various compressed gases used in healthcare would be an example of such a hazard. The construction of a building will also be an important consideration in terms of its susceptibility to the start and spread of a fire. Both building materials and building design are important factors to consider. Similarly fire detection and alarm systems designed to provide timely notification to building occupants in the event of a fire and the sprinkler and/or standpipe systems designed for use in fire control are important fire protection attributes of building design. While standpipe hoses are designed for firefighter use during firefighting operations, automatic sprinkler systems are designed to self-activate, control, and ideally extinguish a fire without human intervention. An additional consideration with building fires is exposures, or buildings that are within close proximity to a building that is on fire and are susceptible to fire spread from the original fire building. The aforementioned information on a building can be collected, analyzed, and summarized in a preplan in advance on an emergency incident, such as a fire in the building.

Incident Priorities

As with all incidents, the priorities in the management of a building fire, in priority order, are (1) life safety, (2) incident stabilization, and (3) property conservation.

Operational Overview

Review of the various types of building fires to which a fire department may be dispatched reveals that while there will be some common practices in managing building fires, the specific tactics utilized will be driven by the aforementioned incident priorities and based on the specific needs of an incident. The same will be true for the safety provisions implemented at a particular incident. The process of incident management delineated in chapter 8 begins with the incident commander performing an initial size-up of the incident upon which strategic goals, tactical objectives, and resource assignments will be based. In cases where a preplan has been prepared and is available, the incident commander will utilize this information. Included in the size-up will be a determination of the dimensions of the building, the number of stories of the building, the building occupancy, the nature of the incident, the specific location and extent of the fire, and whether people are still in the building. The time of the day will also be important in that the census of certain buildings, such as businesses, will typically be reduced at night, whereas the opposite would usually be the case with residences. Fires in basements (cellars), attics, chimneys, and garages can present unique issues, as can fires in high-rise buildings, whether occupied by residential or office units. An important consideration in the event of a building fire will be whether the building has sprinklers and/or standpipes and whether there are fire hydrants within close proximity to the building from which the fire department can secure a necessary water source rather than having to lay extended supply lines or to transport water to the fire scene through the use of water tankers (tenders). As always, life safety will take precedence and will determine the initial and subsequent actions of fire and emergency service personnel, including law enforcement officers who will often assist with the building evacuation and accounting for building occupants.

Based on the initial size-up and continuing size-up activities throughout the incident, the incident commander will formulate appropriate strategic goals and tactical objectives, and assign the various tactics to appropriate resources, such as engine companies, ladder companies, rescue companies, and emergency medical units. Firefighting crews will be assigned the task of conducting search and rescue for building occupants and for ensuring their safe evacuation from the building. Crews will be assigned to locate, confine, and extinguish the fire. Appropriate ventilation techniques will be utilized to facilitate firefighting operations. Exposures will be protected through the application of water to prevent fire spread. Firefighting crews will likewise integrate appropriate overhaul and salvage into their firefighting activities. Building utilities must be controlled in the interest of life safety and utilized as appropriate to support firefighting operations, an example being the use of an HVAC system in performing building ventilation. The fire attack may be conducted as an offensive attack, whereby firefighting personnel are committed to the building, or a defensive attack, where master streams in the form of deck guns and ladder pipes are utilized to protect exposures and to extinguish the fire (fig. 10–3).

Fig. 10–3. Defensive attack at a building fire

A number of the aforementioned firefighting operations require that firefighters have handlines capable of delivering the necessary pressure and volume of water to support effective, efficient, and safe firefighting operations. This requires establishing a water supply that is both reliable and sufficient to meet the requirements of an incident. These water supply requirements are usually addressed through initially utilizing the tank water carried on certain apparatus and laying supply lines from nearby hydrants to supply fire apparatus operating on the incident scene. In rural areas, tanker shuttles may be necessary to transport water from remote locations to the incident scene. In cases where a building has a sprinkler and/or standpipe system, the fire department will assign an engine company to supplement the water supply to these systems through connecting to the building's fire department connection. While hydrant systems are obviously advantageous resources when fighting a building fire, there are times when the fire department may experience water supply problems due to malfunctioning hydrants, broken water mains, or inadequate pressure or volume due to the size or condition of a water main. It is also possible that the fire department may experience problems with building fire protection systems, as in the case of the fire at One Meridian Plaza in which three Philadelphia firefighters perished. Most firefighting and rescue operations at building fires involve laddering the building using ground ladders and/or aerial devices. The height of a building will often determine the appropriate approach to laddering a building. High-rise buildings present additional challenges when their height exceeds the capabilities of both ground and aerial ladders, and firefighters must transport themselves, their equipment, and at times water supply lines to upper floors through building stairways.

Safety Considerations

Life safety of firefighting personnel is addressed through the appointment of an incident safety officer (ISO) and the use of a personnel accountability system. Both initially and throughout the incident the incident commander and the ISO will monitor changing conditions in the interest of ensuring firefighter safety. It is possible that the building may begin to lose its integrity and the decision will be made to pull all personnel out of the building and transition from an interior (offensive) attack to an exterior (defensive) attack. While a proactive decision to change attack modes is essential in certain instances to ensure life safety, the conditions within a fire building can rapidly degrade and in such cases an emergency evacuation may be necessary. General safety concerns associated with building fires include placing ladders in contact with charged electrical lines; the occurrence of a backdraft or flashover; a building collapse; dangers associated with lightweight construction; and firefighters becoming lost, disoriented, trapped, or running out of breathing air. Personal alert safety system (PASS) devices are essential in locating and rescuing downed firefighters. Thermal imaging cameras (TICs) are likewise invaluable in locating victims. One or more rapid intervention teams (RITs) is usually assigned to address such unfortunate eventualities. The physical demands of firefighting, particularly interior firefighting, dictate the need for proper firefighter rehabilitation. Throughout the incident the ISO will focus on personnel safety and continuously monitor situational and human factors that could compromise the safety of firefighters and communicate this information to the incident commander. In addition to ensuring the life safety of emergency responders, the incident commander has responsibility for overall scene safety, including ensuring the public's safety. After major incidents, fire and emergency service organizations and their members can gain insights into how they can enhance their future effectiveness, efficiency, and safety through participating in a post-incident analysis (PIA).

Resource Requirements

The typical response to a building fire will include the fire department, emergency medical services, and law enforcement. While the necessary apparatus and equipment will be based on the nature of the incident, the apparatus typically dispatched to building fires will include engine companies, ladder companies, rescue companies, water tankers (in unhydranted areas), and emergency medical units. The typical tools used by firefighters at building fires include attack lines (handlines), air cascade systems, deck guns, generators, ground ladders, hand tools, high-rise packs, positive pressure ventilation fans, smoke ejectors, thermal imaging cameras, and supply hose lines.

Media Coverage of Building Fires

Building fires represent a significant number of the emergency incidents covered by the local news media. Given that these stories are of interest to readers, listeners, or viewers, they likewise represent stories that the news media is interested in covering. On slow news days it is not unusual for all of the building fires within the coverage area to receive news coverage, whereas on busy news days there will likely be limited coverage of these stories, unless they involve major fires at known locations or have a major community impact. Building fires resulting in civilian and/or firefighter injuries and/or fatalities are likely to receive extensive initial and often follow-up coverage. Specific guidance with respect to the media coverage of incidents involving civilian injuries or fatalities is provided in chapter 21, with companion guidance on media coverage of firefighter injuries and line-of-duty deaths provided in chapter 22.

Building fires, particularly those where photojournalists on the ground or in news helicopters can capture and disseminate vivid images of the fire through live video, are of particular interest to the local news media. Often such coverage subsequently gets picked up by their affiliated networks. Video images captured by the public and sent to news stations are appearing with growing frequency in the news coverage of building and other fires. The media will always be interested in getting and reporting the story in an accurate, comprehensive, professional, and timely manner. The appointment of a PIO provides a single point of contact who will be able to disseminate information or make arrangements for news reporters to interview appropriate individuals, such as the incident commander or an authorized representative from law enforcement or the fire marshal's office who is able to discuss the origin and cause of the fire. Media coverage of arson fire incidents is discussed in chapter 23.

As discussed previously, it will be in the best interest of all involved—fire and emergency service organizations, the news media, and the public—to make sure that reported stories only incorporate information that is factual and credible. While the news media will often interview members of the public, such as family members or neighbors at a residential fire, individuals who were in a store or office building when a fire occurred, or members of the general public who witnessed the incident, it is imperative to ensure that the accurate information provided by official sources serves as the foundation for the story. It is important to remember that the news media has a job to do—to get and report the story—and that if official sources do not make themselves available, reporters may be forced to talk to members of the public who are less well informed about the incident. This is especially important in cases of injuries and/or fatalities. The cautions discussed in chapters 21 and 22 regarding the release of personal information are of utmost importance in professional news coverage of these tragic events.

Given that building fires have the potential of resulting in road closures, detours, and business or other service disruptions, fire and emergency service organizations often seek the assistance of the news media in communicating pertinent information to the community. Information on whether building occupants have been evacuated to another location or are being sheltered in place will likewise be of crucial interest to some members of the public, such as the parents of a child enrolled in the school at which a fire occurred or the family of an elderly relative living in an assisted living facility that experienced a fire.

The job aids provided in figures 10–4 and 10–5 are designed to serve as resources for fire department and media personnel. Figure 10–4 provides an overview of incident management of building fires, while figure 10–5 provides guidance with respect to media coverage of these incidents.

	Incident Management: Building Fires
Typical Incidents	• Assembly occupancies (amusement arcade, exhibit hall, house of worship, movie theatre, nightclub, restaurant, showplace, or sports arena) • Business occupancies (offices) • Educational occupancies (elementary or secondary school) • Industrial occupancies (factory or processing plant) • Institutional occupancies (ambulatory healthcare facility, boarding house, correctional facility, day care center, detention center, healthcare facility, hospital, or nursing home) • Mercantile occupancies (retail store, shopping mall, or strip shopping center) • Residential occupancies (apartment building, condominium, dormitory, hotel, mobile home, motel, rooming house, row house, single-family residence, townhouse, or multi-family residence [e.g., duplex or triplex]) • Storage occupancies (bulk storage facility, distribution center, garage, storage facility, or warehouse) • Miscellaneous occupancies (portable buildings) • *Fire location (attic, basement, chimney, high-rise, or roof)*
Incident Priorities	1. Life safety 2. Incident stabilization 3. Property conservation
Operational Tactics*	• Confinement, crowd control, evacuation, exposure protection, extinguishment, fire protection system support, forcible entry, mechanical systems control, overhaul, preplan review, rehabilitation, salvage, scene assessment, scene management, search and rescue, selection of appropriate operational mode, size-up, staging, traffic control, utility control, ventilation, and water supply
Safety Considerations*	• Backdraft, building collapse, building construction, building hazards, firefighting hazards, flashover, hazardous materials, injuries, lightweight construction, mayday (becoming disoriented, lost, trapped, or running out of breathing air), personnel accountability, rehabilitation, and utilities
Responding Agencies*	• Fire department, emergency medical services, and law enforcement
Resource Requirements	Personnel* • Fire department, emergency medical services, and law enforcement

Fig. 10–4. Incident management job aid. **Note:* The nature of a particular incident will determine the appropriate operational tactics, safety considerations, responding agencies, and resource requirements.

	Personal protective equipment* • Coat, pants, boots, helmet, gloves, eye protection, protective hood, personal alert safety system (PASS), and self-contained breathing apparatus (SCBA) Apparatus* • Aerial apparatus, firefighting apparatus, rescue apparatus, water supply apparatus, and emergency medical units Equipment* • Attack lines (handlines), air cascade system, deck guns, ground ladders, hand tools, high-rise packs, portable equipment, positive pressure ventilation fans, smoke ejectors, supply hose, and thermal imaging cameras

Fig. 10–4. Incident management job aid (continued)

Media Coverage: Building Fires	
WHO?	• Building owner? • Building occupants? • Number of victims? • Number of residents displaced? • Number of and extent of civilian injuries? • Number of and cause of civilian fatalities? • Number of and extent of firefighter injuries? • Number of and cause of firefighter fatalities? • Number of missing or unaccounted for persons? • Personal information on injured individuals? • Personal information on fatalities? • Personal information on missing persons? • *Have appropriate notifications been made?* • *Has information on victims been received from a credible and authorized source?* • *Has information been verified for accuracy?* • How and to which medical facilities were victims transported? • Medical condition of victims (initial and present)? • *See chapter 21 for incidents involving civilian injuries or fatalities.* • *See chapter 22 for incidents involving firefighter injuries or fatalities.*
WHAT?	• Occupancy type? • Specific building use? • Mixed occupancy? • Business name? • Other businesses housed in building?

Fig. 10–5. Media coverage job aid

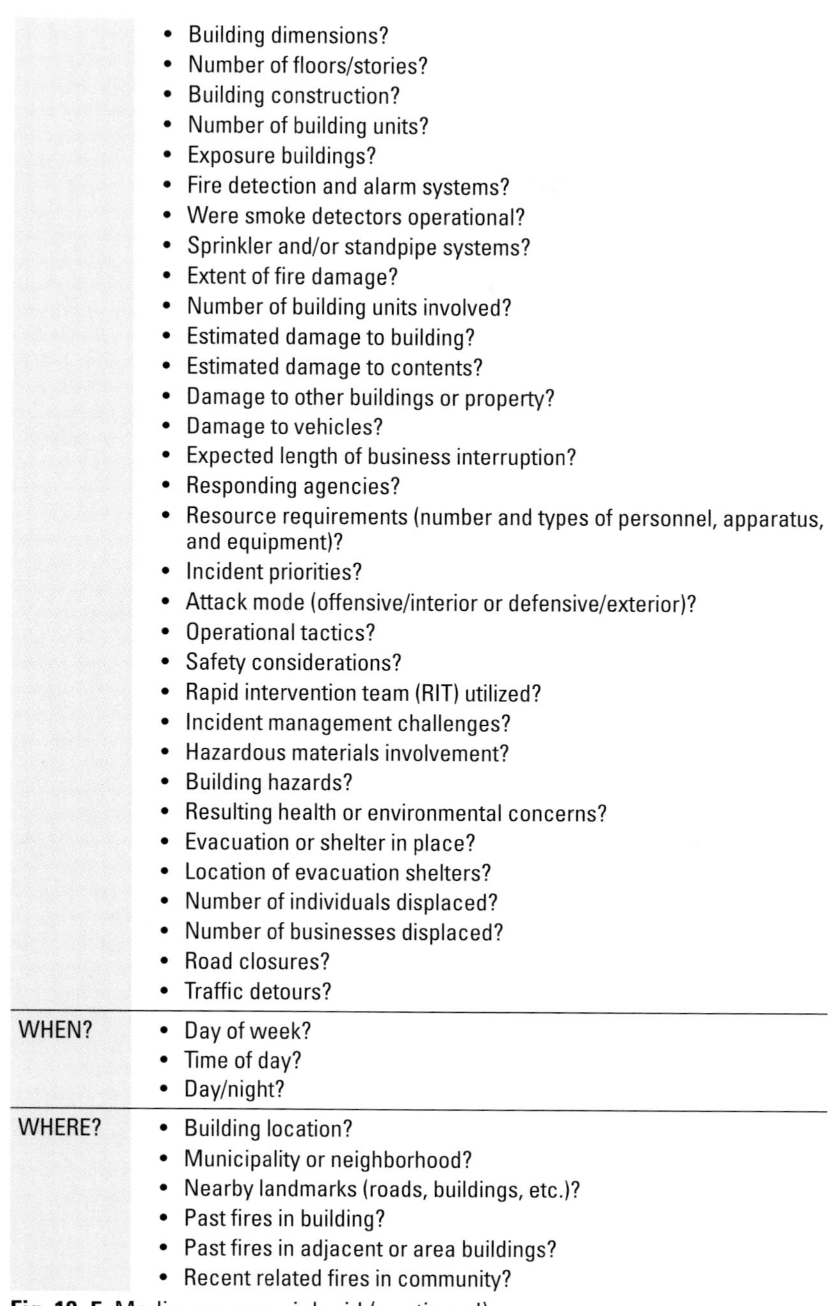

	• Building dimensions? • Number of floors/stories? • Building construction? • Number of building units? • Exposure buildings? • Fire detection and alarm systems? • Were smoke detectors operational? • Sprinkler and/or standpipe systems? • Extent of fire damage? • Number of building units involved? • Estimated damage to building? • Estimated damage to contents? • Damage to other buildings or property? • Damage to vehicles? • Expected length of business interruption? • Responding agencies? • Resource requirements (number and types of personnel, apparatus, and equipment)? • Incident priorities? • Attack mode (offensive/interior or defensive/exterior)? • Operational tactics? • Safety considerations? • Rapid intervention team (RIT) utilized? • Incident management challenges? • Hazardous materials involvement? • Building hazards? • Resulting health or environmental concerns? • Evacuation or shelter in place? • Location of evacuation shelters? • Number of individuals displaced? • Number of businesses displaced? • Road closures? • Traffic detours?
WHEN?	• Day of week? • Time of day? • Day/night?
WHERE?	• Building location? • Municipality or neighborhood? • Nearby landmarks (roads, buildings, etc.)? • Past fires in building? • Past fires in adjacent or area buildings? • Recent related fires in community?

Fig. 10–5. Media coverage job aid (continued)

	• Public water supply (hydrants)? • Source of water supply? • Water utility? • Electric utility? • Gas utility?
WHY?	• Factors that contributed to the fire? • Factors that caused the fire to grow and spread? • Contributing human factors? • Weather-related factors? • Problems with building fire protection systems? • Adequacy of water supply? • Problems with water supply? • Building or fire code violations?
HOW?	• How did the fire start? • Area where the fire started (area of origin)? • Official determination of fire cause(s)? • Is the fire being investigated? • Agencies involved in fire investigation? • Are charges being brought against responsible parties? • *See chapter 23 for information on arson and incendiary fires.*
Sidebar Story Ideas	• Alternative heater safety • Appliance fires • Appliance safety • Arson • Campus fire safety • Candle safety • Careless smoking • Children and fires • Cooking fires • Electrical fires • Fire extinguishers • Fire safety for individuals with disabilities • Fire safety for the elderly • Fire safety for the hearing impaired • Fire safety for the visually impaired • Holiday fire safety • Home exit plans • Home fire hazards • Home fire safety • Incendiary fires • Juvenile firesetting • Residential fire fatalities

Fig. 10–5. Media coverage job aid (continued)

	• Residential sprinklers • Smoke alarms • Winter fires
Resources	• National Fire Protection Association (NFPA) www.nfpa.org • United States Fire Administration (USFA) www.usfa.fema.gov

Fig. 10–5. Media coverage job aid (continued)

Chapter Questions

1. Identify the typical categories of building fires.
2. List the incident management priorities in managing building fires.
3. Discuss several operational tactics related to building fires.
4. Identify the safety concerns related to building fires.
5. Identify the emergency service and support organizations and agencies that are dispatched to building fires.
6. Discuss the resource requirements, in terms of personnel, apparatus, and equipment, associated with building fires.
7. Relate and explain the essential story elements of news coverage of building fires.

Notes

1 National Fire Protection Association. 2012. *National Fire Protection Association Estimates.* Quincy, MA: National Fire Protection Association.

Chapter 11
Equipment Fires

Chapter Objectives

- Identify the typical categories of equipment fires.
- Identify the incident management priorities for equipment fires.
- Discuss the operational tactics utilized at equipment fires.
- Discuss the safety considerations associated with equipment fires.
- Identify the emergency service agencies that respond to equipment fires.
- Identify the resource requirements associated with equipment fires.
- Discuss the story elements associated with media coverage of equipment fires.

Many of the necessities and conveniences of modern society are made possible through the advent and commercialization of electricity, as well as the harnessing and distribution of other natural resources including various gases. This chapter considers fires involving various types of residential, commercial, and industrial equipment. Consideration of the issues associated with infrastructure emergencies are the focus of chapter 16. While equipment fires originate from a number of sources, typical factors that contribute to the ignition of an equipment fire include electrical failure or malfunction; mechanical failure or malfunction; deficiency in equipment design, manufacture, or installation; misuse of equipment; and operational deficiency. While certain equipment fires occur as a consequence of situational factors, such as overheated equipment or electrical shorts, human factors in the use of equipment likewise contribute to equipment fires, as in the case of the improper use of portable space heaters in a residence or business enterprise.

Typical Incidents

Equipment fires occur in all of the types of occupancies discussed in chapter 10. While these fires can be self-contained to the particular piece of equipment involved, as listed in figure 11–1, they often spread to exposures and result in a building and/or contents fire. This would be the case with a fire that started in a portable heating unit and spread first to a chair and blanket and subsequently to other contents and the involved room within a residence, or to nearby boxes in storage and subsequently to the building in an office, factory, or distribution center. Thus, many fires that originate and are dispatched as equipment fires may in reality extend to exposures and become a building fire by the time the fire department arrives.

- Appliance fire
- Boiler fire
- Compactor fire
- Cooking fire
- Electrical equipment fire
- Heater or furnace fire
- Incinerator fire
- Outside equipment fire
- Portable heater fire

Fig. 11–1. Typical incidents: equipment fires

The equipment involved in a fire typically operates on a power source that can contribute to the fire, as well as dictate the appropriate approach to extinguishing the fire. Primary power sources include electricity and various gases, including gasoline, diesel fuel, propane, and natural gas. The properties of the various hazardous materials used as power sources, as well as appropriate operational modes and tactics, are discussed in chapters 12 and 15. This chapter primarily focuses on issues with respect to fires involving electrical equipment. The major hazard involved with electrical equipment, regardless of whether on fire, is obviously the potential of personnel coming in contact with charged electrical circuits or equipment, resulting in potential injuries and/or death.

Heating and cooking fires comprise two major categories of residential equipment fires. Heating fires occur in both permanently installed equipment such as a heater or furnace and in portable or alternative heating units such as a portable space heater. Fires involving cooking equipment involve appliances such as electric or gas stoves and microwave ovens. A contributing factor to many types of equipment fires in a residence can be human behavior, as in the case of unattended cooking or placing a portable heater in close proximity to other home furnishings. The many other appliances contained within a modern residence also present the potential for equipment fires. In addition to the many electrical appliances that derive their source of power from a residence's electrical service, the electrical circuits and equipment also present the potential for electrical fires.

Equipment fires in other occupancies can likewise include those discussed with respect to residences, including appliance fires, heater or furnace fires, and portable heater fires. The nature of these buildings will often determine the specific equipment incorporated into their operations and facilities, such as the equipment utilized in operational processes. Common equipment utilized in nonresidential occupancies may result in boiler fires, compactor fires, electrical equipment fires, and incinerator fires. Equipment fires can occur within or outside a building. The particular equipment that is present and susceptible to fire will vary greatly in accordance with the nature of an occupancy and its associated business processes.

As diverse as this equipment may be, it can usually be divided into the following categories: commercial and medical equipment; electrical distribution, lighting, and power transfer equipment; electronic and other electrical equipment; garden tools and agricultural equipment; heating, ventilation, and air conditioning equipment; kitchen and cooking equipment; personal and household equipment; and shop tools and industrial equipment (fig. 11–2). Each particular category of equipment and specific equipment and its use present unique issues, and at times challenges, with respect to firefighting efforts.

Fig. 11–2. Equipment in a commercial kitchen

Incident Priorities

As with all incidents, the priorities in the management of an equipment fire, in priority order, are (1) life safety, (2) incident stabilization, and (3) property conservation.

Operational Overview

While some equipment fires occur outside of buildings or structures, as in the case of a fire involving a generator designed to power a facility in the event of a community power outage, others occur in close proximity to a building as in the case of a trash compactor fire, and most equipment is actually resident within buildings. This is an extremely important consideration from the standpoint of the potential for a fire to extend from the involved equipment to exposed property, including the building and its contents. The issue of fire confinement and exposure protection is particularly important in the management of an equipment fire.

Although protecting exposures and thus preventing the spread of the fire is obviously important, the most important priority in handing any type of equipment fire is ensuring the life safety of the public and firefighters. The power sources from which modern equipment derive their functionality present inherent hazards to emergency responders and others potentially exposed to them. In the case of electrical equipment, there is always the potential for personnel to come in contact with energized electrical circuits, which can result in injuries or fatalities from electrocution. Firefighting activities involving fires in electrical equipment should only be undertaken when it can be assured and confirmed that all involved equipment has been properly de-energized. Appropriate facilities personnel and electric company representatives are often necessary to ensure the safety of these procedures (fig. 11–3).

Fig. 11–3. Facility personnel de-energizing electrical circuits. (Courtesy of M. Grant Everhart.)

Other equipment power sources can likewise present hazards to firefighters and the public based on the properties of these hazardous materials. These materials may be flammable, combustible, or potentially explosive and should be appropriately managed in all firefighting operations. As with electrical equipment fires, shutting off the power source, in this case a gas line feeding the involved equipment, is an essential element of effective, efficient, and safe firefighting operations. Additional information on the management of hazardous materials fires and incidents is found in chapters 12 and 15, respectively.

The application of an appropriate extinguishing agent is of paramount importance when handling equipment fires. The selection of an appropriate extinguishing agent relates to the three incident priorities articulated above. The inappropriate application of water to energized electrical equipment would compromise the top priority of life safety in that the water applied in the interest of extinguishing the fire could readily conduct electricity, posing the threat of electrocution. The application of water as an extinguishing agent in the presence of any water-reactive materials could likewise compromise firefighter life safety. The selection and application of the proper extinguishing agent supports the priority of stabilizing the incident in an effective, efficient, and safe manner. Given that the application of the wrong extinguishing agent can further damage involved equipment, particularly computer equipment and other electronic devices, the application of an appropriate extinguishing agent also contributes to the third priority in incident management—property conservation.

The operational mode in the management of an equipment fire will therefore begin with ensuring the life safety of firefighters by properly de-energizing electrical circuits and/or shutting off other utility feeds or lines. Utility control remains important throughout the management of the incident in the interest of ensuring that no one intervenes by turning the power or fuel source back on. When available, preplans should be reviewed for important facility information, including the location of electrical and gas shutoffs. Consulting with knowledgeable facility personnel is always prudent in these instances. In addition to controlling utilities, firefighting personnel should support fire protection systems and control the building's mechanical systems, including HVAC systems, as appropriate and necessary to support effective, efficient, and safe firefighting operations.

Firefighting tactics will involve the confinement and extinguishment of the equipment fire and any extension of the fire to the building or its contents. A determination must be made regarding an appropriate extinguishing agent and application technique. Exposure protection is an important element of firefighting of equipment fires. Exposures in a factory or processing plant could include adjacent, often integrated, equipment, raw material stocks, and finished products. Successful incident management of equipment fires will often include appropriate ventilation designed to remove smoke and other products of combustion, overhaul activities designed to ensure that the fire is fully extinguished, and salvage activities designed to protect equipment and other property from further damage, including that possibly resulting from the application of extinguishing agents during fire suppression. Thermal imaging cameras are useful in checking for "hot spots" in equipment and surrounding property. Evacuation of all but essential firefighting personnel from the area surrounding the fire should be handled early in the incident in the interest of life safety. Likewise, accountability for all firefighting personnel must be ensured throughout the incident, and appropriate decontamination of firefighting personnel should take place following any related exposure.

Safety Considerations

While equipment fires vary widely from a seemingly simple appliance fire in a home to an extremely complex fire in the equipment of a manufacturing plant or processing facility or in the lighting or broadcast equipment in a television studio, certain inherent risks, such as the potential for electrocution, present themselves at these incidents. While the capacity and complexity of the electrical systems in different occupancies will vary significantly, the potential for harm resulting from electrocution must always drive the decisions of the incident commander and other firefighting personnel operating on the scene of an equipment fire.

The presence and challenges of hazardous materials must also be determined and addressed in the safe management of these emergency incidents. Air quality monitoring should be conducted throughout these incidents. The critical importance of having knowledgeable personnel deenergize electrical circuits and/or shut off fuel supplies bears repeating, as does the importance of ensuring that these sources are not reactivated during the management of the incident. In addition to the aforementioned hazards, the potential can exist for firefighters to suffer serious injuries if involved mechanical equipment is still operational.

As always, the presence of hazardous materials should be anticipated and properly addressed throughout the management of the incident. Equipment fires have the potential to produce harmful atmospheres, thus demanding that all personnel operating in proximity to equipment fires don appropriate personal protective equipment including respiratory protection. In cases where firefighters become contaminated during firefighting operations, appropriate decontamination procedures must be implemented. An additional challenge of equipment fires is that the equipment is often located in a confined space and thus provides limited access and/or ventilation.

The selection of an appropriate extinguishing agent based on the involved equipment and situation provides the foundation of effective, efficient, and safe incident management of equipment fires. Given the inherent hazards associated with equipment fires, careful supervision as well as personnel accountability must be maintained throughout these incidents. The essential role of an ISO at these incidents should be obvious.

Resource Requirements

The typical response to an equipment fire, unless extension to a building is reported, is often limited to the fire department and perhaps law enforcement. While the necessary apparatus and equipment will be based on the nature of the incident, the apparatus typically dispatched to equipment fires will include engine companies, with ladder companies or rescue companies being dispatched based on specific incident needs. The typical tools used by firefighters at equipment fires include extinguishing agents, fire extinguishers, gas detectors, generators, hand tools, positive pressure ventilation fans, smoke ejectors, and thermal imaging cameras.

Media Coverage of Equipment Fires

While many equipment fires will prove to be of limited, if any, interest to the public and the news media, those that produce more significant consequences will typically appear on the news media's radar screen in terms of being a newsworthy story to cover. A fire that started as an equipment fire and spread to the building, causing extensive damage would usually generate increased interest on the part of the public and the news media. Regardless of the extent of the fire, were it to result in casualties in terms of civilian and/or firefighter injuries and/or fatalities, it would in all likelihood be deemed worthy of news coverage. Were the equipment fire to result in a business interruption negatively impacting the public, it would also be likely to receive news coverage, as in the case of an equipment fire in a postal processing and distribution center or an anchor store in a regional shopping mall during a busy holiday mailing or shopping season.

The news coverage of equipment fires is therefore somewhat different than the news coverage of certain other types of emergency incidents. Typically, the equipment fire as such may be deemed fairly inconsequential, with the real focus of the news coverage being on the impact of that fire in terms of its consequences. While it would be rather unusual for an equipment fire contained to a portable heater being used in a residential unit in an apartment complex to receive news coverage, were that same fire to result in serious injuries and/or fatalities or in major fire damage displacing the residents of a dozen residential units, it would likely become a breaking news story. It should be noted that these fires also afford the news media an opportunity to provide valuable public education information on a range of topics such as the appropriate use of portable heaters. In cases where the fire occurred in a nonresidential occupancy and yielded a significant business impact such as those described above, the news media can provide valuable information to the public, for example, the impact of the fire at the postal facility on their mail service, or the stores that are open or closed in the shopping mall as a result of the fire. These are additional examples of the collaborative dissemination of information to that public that is discussed and advocated throughout this book.

The job aids in figures 11–4 and 11–5 are designed to serve as resources for fire department and media personnel. Figure 11–4 provides an overview of incident management of equipment fires, while figure 11–5 provides guidance with respect to media coverage of these incidents.

Incident Management: Equipment Fires	
Typical Incidents	• Appliance fire • Boiler fire • Compactor fire • Cooking fire • Electrical equipment fire • Heater or furnace fire • Incinerator fire • Outside equipment fire • Portable heater fire
Incident Priorities	1. Life safety 2. Incident stabilization 3. Property conservation
Operational Tactics*	• Confinement, decontamination, evacuation, exposure protection, extinguishment, fire protection system support, mechanical systems control, overhaul, personnel accountability, review preplan, salvage, scene security, utility control, and ventilation
Safety Considerations*	• Appropriate extinguishing agent and technique, carbon monoxide, confined space (limited access, limited ventilation, potentially harmful atmosphere), contamination, electrocution, equipment electrical hazards, equipment mechanical hazards, evacuation, exposure, hazardous materials, personnel accountability, respiratory protection, toxic atmospheres, utilities, victims, and water-reactive materials
Responding Agencies*	• Fire department and facility plant personnel (electrician)
Resource Requirements	Personnel* • Fire department and emergency medical services (in the event of injuries) Personal protective equipment* • Coat, pants, boots, helmet, gloves, eye protection, protective hood, personal alert safety system (PASS), and self-contained breathing apparatus (SCBA) Apparatus* • Firefighting apparatus Equipment* • Building fire protection systems, extinguishing agents, fire extinguishers, gas detectors, generators, hand tools, positive pressure ventilation fans, smoke ejectors, and thermal imaging cameras

Fig. 11–4. Incident management job aid. **Note:* The nature of a particular incident will determine the appropriate operational tactics, safety considerations, responding agencies, and resource requirements.

Media Coverage: Equipment Fires	
WHO?	• Property owner? • Number of victims? • Number of and extent of civilian injuries? • Number of and cause of civilian fatalities? • Number of and extent of firefighter injuries? • Number of and cause of firefighter fatalities? • Personal information on injured individuals? • Personal information on fatalities? • *Have appropriate notifications been made?* • *Has information on victims been received from a credible and authorized source?* • *Has information been verified for accuracy?* • How and to which medical facilities were victims transported? • Medical condition of victims (initial and present)? • *See Chapter 21 for incidents involving civilian injuries or fatalities.* • *See Chapter 22 for incidents involving firefighter injuries or fatalities.*
WHAT?	• Involved equipment? • Occupancy type? • Specific building use? • Business name? • Other businesses housed in building? • Extent of fire damage? • Estimated damage to equipment? • Estimated damage to building? • Estimated damage to contents? • Damage to other buildings or property? • Expected length of business interruption? • Responding agencies? • Resource requirements (number and types of personnel, apparatus, and equipment)? • Incident priorities? • Operational tactics? • Safety considerations? • Hazardous materials involvement? • Equipment hazards? • Resulting health or environmental concerns?
WHEN?	• Day of week? • Time of day? • Day/night?

Fig. 11–5. Media coverage job aid

WHERE?	• Incident location? • Municipality or neighborhood?
WHY?	• Factors that contributed to the fire? • Factors that caused the fire to grow and spread? • Contributing human factors? • Problems with building fire protection systems? • Building or fire code violations?
HOW?	• How did the fire start? • Area where the fire started (area of origin)? • Official determination of fire cause(s)? • Is the fire being investigated? • Agencies involved in fire investigation? • *See chapter 23 for information on arson and incendiary fires.*
Sidebar Story Ideas	• Alternative heater safety • Appliance fires • Appliance safety • Cooking fires • Electrical fires • Electrical fire safety • Heater/furnace fires • Holiday fire safety • Smoke alarms
Resources	• Consumer Products Safety Commission (CPSC) www.cpsc.gov • National Fire Protection Association (NFPA) www.nfpa.org • Occupational Safety and Health Administration (OSHA) www.osha.gov • United States Fire Administration (USFA) www.usfa.fema.gov

Fig. 11–5. Media coverage job aid (continued)

Chapter Questions

1. Identify the typical categories of equipment fires.
2. List the incident management priorities in managing equipment fires.
3. Discuss several operational tactics related to equipment fires.
4. Identify the safety concerns related to equipment fires.
5. Identify the emergency service and support organizations and agencies that are dispatched to equipment fires.
6. Discuss the resource requirements, in terms of personnel, apparatus, and equipment, associated with equipment fires.
7. Relate and explain the essential story elements of news coverage of equipment fires.

Chapter 12
Hazardous Materials Fires

Chapter Objectives

- Identify the typical categories of hazardous materials fires.
- Identify the incident management priorities for hazardous materials fires.
- Discuss the operational tactics utilized at hazardous materials fires.
- Discuss the safety considerations associated with hazardous materials fires.
- Identify the emergency service agencies that respond to hazardous materials fires.
- Identify the resource requirements associated with hazardous materials fires.
- Discuss the story elements associated with media coverage of hazardous materials fires.

The incorporation of hazardous materials into most aspects of everyday life creates the potential for the presence and involvement of hazardous materials in an ever-increasing array of incidents to which a contemporary fire department may respond. While many of these hazardous materials incidents may involve leaks, spills, or other releases without the presence of fire (see chapter 15), the potential for fire often exists. Fires that involve hazardous materials present inherent challenges beyond those normally associated with a particular type of fire.

While this chapter is devoted to an examination of hazardous materials fires, it is important to recognize that the potential for hazardous materials involvement including fire exists in many of the other categories of incidents examined in this book. Various types of accidents, particularly those involving transportation, may involve hazardous materials that potentially can experience ignition and a resulting fire. This could involve the fuel tank of a passenger vehicle or a road tanker transporting fuel to a service station. Vehicle fires can also occur in the absence of an accident, as when a heat source such as a catalytic converter, overheated brakes, or an electrical short causes a fire that subsequently involves onboard hazardous materials. Building fires may likewise possess the potential for hazardous materials fires, ranging from the involvement of a fairly small quantity of cleaning materials in a home to the significant quantities of chemicals utilized within commercial or industrial processes. Equipment fires and infrastructure emergencies likewise present the potential for fire, whether in terms of hazardous materials used within this equipment, such as transformer oil, or the ignition of a leaking or ruptured gas main. Outdoor fires also have the potential for hazardous materials involvement as in the case where a brush or wildland fire impinges upon an outside propane storage tank.

Typical Incidents

While there may be times that a fire department is dispatched to a "hazardous materials fire," in the majority of instances it will instead be dispatched to one of the various types of fires discussed elsewhere in this book—building fires, equipment fires, outside fires, or vehicle fires. Based on their preparation in terms of training, experience, preplanning, and awareness of their response district, fire department personnel will always be alert to the potential for the presence and involvement of hazardous materials.

Hazardous materials fires will typically occur in a fixed facility or in a transportation incident, as listed in figure 12–1. Representative fires at fixed facilities range from a home propane grill or its supply tank on fire to a much larger propane tank involved in a fire at an industrial occupancy. Equipment fires at a fixed facility can also involve hazardous materials used in processing activities. Fires can also involve the processing and storage of hazardous materials, as in the case of a fire in a chemical plant or a refinery.

- Chemical plant fire
- Gasoline fire
- LPG tank fire
- Refinery fire
- Tar fire
- Hazardous materials fire (fixed facility)
- Hazardous materials fire (rail tanker)
- Hazardous materials fire (road tanker)

Fig. 12–1. Typical incidents: hazardous materials fires

Transportation incidents involving hazardous materials fires typically involve rail and road tankers engaged in the transport of hazardous materials (fig. 12–2). These conveyances must by regulation be properly identified with placards and appropriate hazard information. Fires involving hazardous materials in transit can also occur in box trucks and trailers. While many hazardous materials are shipped by rail and road, others are continuously transported through pipelines, whether in major distribution lines or the smaller infrastructure that provides residential and commercial gas service.

Fig. 12–2. Transportation fire involving hazardous materials. (Courtesy of Chester County Hazardous Materials Team.)

Although a fire department will routinely respond to a fairly limited array of hazardous materials incidents, including those involving fire, given the extensive presence of hazardous materials in today's world, it is imperative that fire departments and organizations that provide technical support be prepared to respond to all hazardous materials incidents and address them in a manner consistent with recognized life safety, health, and environmental concerns. An incident as routine as a gasoline, liquid petroleum gas (LPG), tar, or tire fire, if not properly managed, can have significant life safety consequences and environmental impacts. As with all incidents involving hazardous materials, the essential first step in incident management is to determine the involved material(s); their properties in terms of such essential aspects as their potential to bring harm based on flammability, explosion, and health effects; and the appropriate approach that should be taken to effectively, efficiently, and safely resolve the situation in accordance with established incident priorities. More information on the management of hazardous materials incidents is presented in chapter 15.

Incident Priorities

As with all incidents, the priorities in the management of an incident involving a hazardous material on fire, in priority order, are (1) life safety, (2) incident stabilization, and (3) property conservation.

Operational Overview

While the fundamentals of fire extinguishment—apply an appropriate extinguishing agent to the fire—are common for all types of fires, the introduction of hazardous materials into a fire environment can significantly complicate matters based on the properties of the involved hazardous material(s). Some hazardous materials present fire hazards based on their flammability or combustibility; others based on their reactivity or explosive potential (fig. 12–3). A thorough initial and continuous size-up is particularly important when an incident involves hazardous materials that are either on fire or represent potential exposures in terms of becoming involved.

Fig. 12–3. Response personnel researching involved hazardous materials. (Courtesy of Chester County Hazardous Materials Team.)

While the properties of the majority of hazardous materials, including how they react to fire, have been well documented, the reactions that derive from the combination of various hazardous materials and their involvement in a fire are less well known. A thorough understanding of the chemicals involved in a fire is essential in the determination and implementation of an appropriate operational mode. As with other fires, either an offensive fire attack or a defensive posture designed to protect exposures may be appropriate. The application of water as an extinguishing agent would be inappropriate as well as dangerous in instances where involved materials are water-reactive. The health and safety of emergency responders and the public, as well as the potential for an undesirable environmental impact, will influence the selection of an appropriate operational mode. In some cases the decision will be to operate in a nonintervention mode and let the fire burn; in other cases it will be to apply appropriate extinguishing agent(s) and use appropriate operational tactics.

In all incidents involving hazardous materials, it is imperative that personnel, apparatus, and equipment be staged both uphill and upwind. Likewise, personnel deployed to perform tactical assignments should always don appropriate personal protective equipment for the hazardous material(s) involved, including self-contained breathing apparatus (SCBA), and should approach the incident from uphill and upwind. All personnel performing tactical assignments at a hazardous materials incident should have appropriate training and certifications corresponding with the nature of their assigned tasks. While the specific strategic goals and tactical objectives of managing an incident will be incident-specific and driven by the incident priorities, it is important never to extinguish a gas main or line fire until such time as the gas supply has been shut off. The three protective control zones—cold zone, warm zone, and hot zone—should be established and enforced at any hazardous materials incident. Determination of the need to protect the public through sheltering in place or evacuations should be made and implemented. The use of air-monitoring equipment will be integral to making effective decisions regarding the protection of emergency responders and the public. Additional information on these and other related topics regarding the effective, efficient, and safe management of hazardous materials incidents is found in chapter 15.

The nature of the incident, including the involved material(s) and involved or exposed buildings, vehicles, and other property will be considered in the selection of appropriate operational tactics, as will such environmental factors as wind direction and speed and the nature of the occupancies downwind from the incident. Whereas ensuring the health and safety of the public poses a significant challenge in the case where the wind is carrying the toxic by-products of combustion toward nearby high-density facilities or residential areas, if the wind was carrying those same harmful gases in a direction that was rural with little habitation, this would present less of an incident management challenge. While evacuations would likely be considered in both situations, the prudent initial action in the first case may be to alert the involved facility or city agency to initiate the necessary actions to, at least temporarily, shelter its residents in place, including the sealing off of openings to the outside and shutting down HVAC systems that draw air from outside the buildings.

Firefighting tactics will typically include confinement; fire extinguishment using appropriate extinguishing agent(s); exposure protection including cooling exposed tanks, containers, vehicles, and buildings; support of fixed fire protection systems; water runoff control; utility and mechanical systems control; ventilation; and the establishment of an adequate water supply. In cases of an event at a fixed facility, facility preplans and community response plans, if available, can provide important guidance for incident management decision making. Both crowd and traffic control, including establishing necessary detours, is important in these incidents. Continual air monitoring throughout these incidents is important, as is ensuring that all personnel and equipment involved in firefighting activities are properly decontaminated during incident termination activities.

Safety Considerations

The inherent hazards and risks of firefighting are increased, often greatly, when hazardous materials are involved in a fire or the potential exists for the fire to involve exposed hazardous materials. As with all incidents, the appointment of a qualified ISO and use of a personnel accountability system are essential. Appropriate actions must be taken to ensure the safety of the public, including initiating appropriate sheltering in place or evacuating. The use of appropriate crowd and traffic control, in addition to protecting the health and well-being of the public, ensures that responder safety is not compromised through misguided or unintentional actions by the public.

While their potential for harm varies widely based on the properties of specific hazardous materials, the potential expected and unexpected behaviors of these materials, particularly in cases where they are exposed to fire, must be both anticipated and respected. In addition to the potential for injuries and fatalities, exposure to hazardous materials can result in long-term effects in terms of acute and chronic illnesses. Exposure protection of responders is provided through utilizing only those personnel with the requisite training and certifications to perform certain tasks and providing them with appropriate personal protective equipment, including respiratory protection based on the involved chemical(s). The routes through which individuals can be exposed include absorption, ingestion, and inhalation. Proper decontamination of all personnel coming in contact with or being exposed to hazardous materials is essential.

As stated earlier, it is imperative that personnel, apparatus, and equipment be staged uphill and upwind from a hazardous materials incident, and that all approach the incident only from uphill and upwind. The earlier caution of not extinguishing a gas line fire before first shutting off the source of gas supply likewise bears repeating. A primary safety concern in these types of fires is the potential for the occurrence of a boiling liquid expanding vapor explosion (BLEVE) wherein the flames impinging on a container housing hazardous materials cause the material inside the tank to expand, which causes the tank to lose its structural integrity and eventually an extremely dramatic and dangerous explosion. The potential of a BLEVE illustrates the importance of protecting exposures, such as cooling railroad tank cars carrying hazardous materials in the event of a brake fire on the involved rail car. The fact that certain hazardous materials are reactive to water and/or may produce toxic byproducts of combustion likewise present real safety concerns when operating at the scene of a hazardous materials fire.

Resource Requirements

The typical response to a fire involving hazardous materials will include the fire department, a hazardous materials team, emergency medical services, and law enforcement. Additional resources in the form of public works, technical specialists, and hazardous materials cleanup contractors may be requested, based on the nature and resource needs of the incident. Governmental agencies having regulatory or enforcement authority and jurisdiction will be notified and may also respond to the incident. While the necessary apparatus and equipment will be based on the nature of the incident, the apparatus typically dispatched to a fire involving hazardous materials will include firefighting apparatus, rescue apparatus, water supply apparatus, specialized apparatus, and emergency medical units. The typical tools used by emergency responders at these incidents include attack lines (handlines), air cascade systems, deck guns, extinguishing agents, foam generation and application equipment, gas detectors and other metering instruments, hand tools, portable equipment, positive pressure ventilation fans, smoke ejectors, supply hose lines, and thermal imaging cameras.

Media Coverage of Hazardous Materials Fires

Incidents that involve hazardous materials on fire are events that have the potential of attracting significant public and media attention. Likewise, there will be interest on the part of the news media to capture and report on such breaking news stories in a timely manner, preferably in advance of competing media outlets. Frequently, such news stories, accompanied by dramatic video images captured by photojournalists from the ground or air, will not only be a lead story locally, but will be incorporated into regional, national, and at times even international news coverage.

While most emergency incidents are dynamic in nature and thus offer a certain degree of uncertainty, perhaps even suspense, for news readers, listeners, and viewers, this is most certainly true in the case of a hazardous materials fire. Just as the incident commander managing the incident will be interested in determining through a thorough process of size-up and forecasting what has happened, what is happening, and what is likely to happen, so too will the news media and the audiences that they serve. The importance of getting the story right in terms of accuracy, comprehensiveness, professionalism, and timeliness is obvious given the real and/or perceived potential for harm to members of the community and their loved ones. Ensuring that all information reported is accurate and is verified through credible sources is a must in the interest of public safety and welfare, as well as in maintaining the reputation of a reporter and his or her news organization for responsible reporting.

As with many other incidents, fire and emergency service officials may solicit the cooperation and assistance of the news media in ensuring that proper information is disseminated to the public throughout the incident. This information could include alerts, instructions, and status updates. Many radio and television stations have positioned themselves as the "go to" source for public information during times of emergency and thus have a vested interest in successfully disseminating vital information on things such as the hazards facing the community, actions that residents should take to protect themselves and their property, and available resources to assist them. Essential parts of this coverage will often include information about sheltering in place or evacuations, road closures and detours, infrastructure interruptions, and air and water quality. These are primary examples of emergency incidents where, through cooperation and collaborative action, fire and emergency service professionals and their news media counterparts can ensure both the effective, efficient, and safe resolution of the emergency situation and its community and environmental aftermath, and the accurate, comprehensive, professional, and timely reporting of the accompanying news story.

The two job aids in figure 12–4 and 12–5 are designed to serve as resources for fire department and media personnel. Figure 12–4 provides an overview of incident management of hazardous materials fires, while figure 12–5 provides guidance with respect to media coverage of these incidents.

Incident Management: Hazardous Materials Fires	
Typical Incidents	• Chemical plant fire • Gasoline fire • LPG tank fire • Refinery fire • Tar fire • Hazardous materials fire (fixed facility) • Hazardous materials fire (rail tanker) • Hazardous materials fire (road tanker)
Incident Priorities	1. Life safety 2. Incident stabilization 3. Property conservation
Operational Tactics*	• Air monitoring, confinement, continuous size-up, cool exposed tanks, crowd control, decontamination, establish control zones (hot, warm, and cold), evacuation, exposure protection, extinguishment, identification of hazardous materials and properties, mechanical systems control, review preplan, rehabilitation, runoff control, scene assessment, scene management, selection of appropriate operational mode, shelter in place, staging, support fixed fire protection systems, traffic control, utility control, ventilation, and water supply
Safety Considerations*	• Approach from uphill, approach from upwind, do not extinguish gas leak fire until gas supply has been shut off, don appropriate personal protective equipment (PPE) including self-contained breathing apparatus (SCBA), identify hazardous material(s) involved and properties, boiling liquid expanding vapor explosion (BLEVE), contamination, control zones, explosion, exposure (absorption, ingestion, or inhalation), evacuation and/or shelter in place, fatalities, hazardous materials, injuries, personnel accountability, respiratory protection, utilities, victims, and water-reactive materials
Responding Agencies*	• Fire department, emergency medical services, law enforcement, hazardous materials team, technical specialists, public works, hazardous materials cleanup contractor, and representatives of governmental agencies having jurisdiction
Resource Requirements	Personnel* • Fire department, emergency medical services, law enforcement, and hazardous materials team

Fig. 12–4. Incident management job aid. **Note:* The nature of a particular incident will determine the appropriate operational tactics, safety considerations, responding agencies, and resource requirements.

	Personal protective equipment* • Coat, pants, boots, helmet, gloves, eye protection, protective hood, specialized chemical protective equipment, personal alert safety system (PASS), and self-contained breathing apparatus (SCBA) Apparatus* • Firefighting apparatus, aerial apparatus, rescue apparatus, water supply apparatus, emergency medical units, and specialized apparatus Equipment* • Attack lines (handlines), air cascade system, deck guns, extinguishing agents, foam generation and application equipment, gas detectors and other monitoring instruments, hand tools, portable equipment, positive pressure ventilation fans, smoke ejectors, supply hose, and thermal imaging cameras

Fig. 12–4. Incident management job aid (continued)

	Media Coverage: Hazardous Materials Fires
WHO?	• Building owner? • Building occupants? • Number of victims? • Number of residents displaced? • Number of and extent of civilian injuries? • Number of and cause of civilian fatalities? • Number of and extent of firefighter injuries? • Number of and cause of firefighter fatalities? • Number of missing or unaccounted for persons? • Personal information on injured individuals? • Personal information on fatalities? • Personal information on missing persons? • *Have appropriate notifications been made?* • *Has information on victims been received from a credible and authorized source?* • *Has information been verified for accuracy?* • How and to which medical facilities were victims transported? • Medical condition of victims (initial and present)? • *See chapter 21 for incidents involving civilian injuries or fatalities.* • *See chapter 22 for incidents involving firefighter injuries or fatalities.*

Fig. 12–5. Media coverage job aid

WHAT? NOTE: Most hazardous materials fires also involve vehicles or buildings.	• Nature of incident? • Hazardous material(s) involved? • Properties of involved hazardous materials? • DOT hazard classification(s)? • Type of incident (fixed, mobile, natural, or portable)? • Type of container? • Fixed facility? • Rail tanker? • Road tanker? • Container capacity? • Estimated release amount or fuel load? • Physical state when released? • Related information? • Buildings involved? • Vehicles involved? • Other property involved? • Exposed buildings? • Other exposed property? • Extent of fire damage? • Estimated damage to building? • Estimated damage to contents? • Damage to other buildings or property? • Damage to vehicles? • Expected length of business interruption? • Responding agencies? • Resource requirements (number and types of personnel, apparatus, and equipment)? • Incident priorities? • Attack mode (offensive, defensive, or nonintervention)? • Operational tactics? • Safety considerations? • Incident management challenges? • Resulting health or environmental concerns? • Evacuation or shelter in place? • Location of evacuation shelters? • Number of individuals displaced? • Number of businesses displaced? • Road closures? • Expected duration of road closures? • Traffic detours? • *See chapter 10 for incidents involving building fires.*

Fig. 12–5. Media coverage job aid (continued)

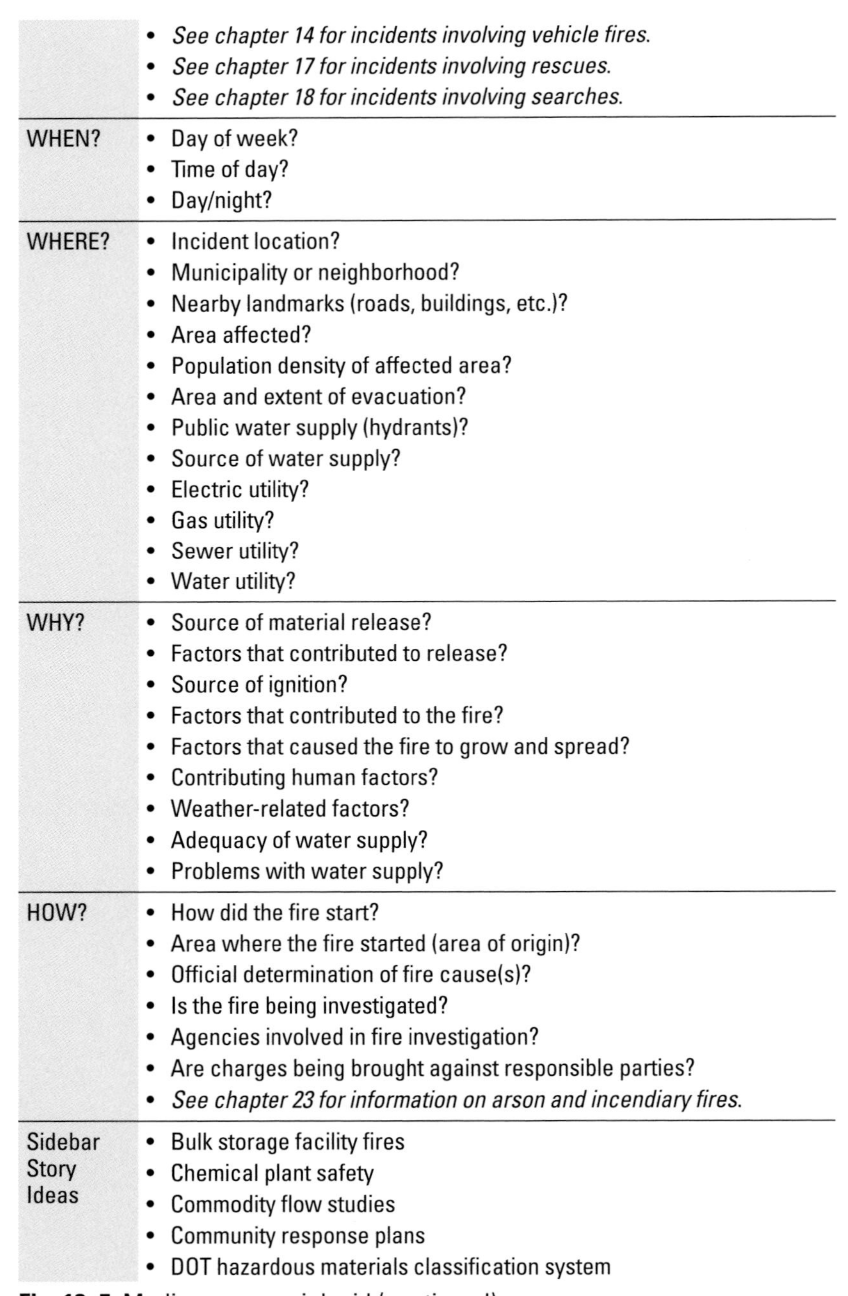

	• *See chapter 14 for incidents involving vehicle fires.* • *See chapter 17 for incidents involving rescues.* • *See chapter 18 for incidents involving searches.*
WHEN?	• Day of week? • Time of day? • Day/night?
WHERE?	• Incident location? • Municipality or neighborhood? • Nearby landmarks (roads, buildings, etc.)? • Area affected? • Population density of affected area? • Area and extent of evacuation? • Public water supply (hydrants)? • Source of water supply? • Electric utility? • Gas utility? • Sewer utility? • Water utility?
WHY?	• Source of material release? • Factors that contributed to release? • Source of ignition? • Factors that contributed to the fire? • Factors that caused the fire to grow and spread? • Contributing human factors? • Weather-related factors? • Adequacy of water supply? • Problems with water supply?
HOW?	• How did the fire start? • Area where the fire started (area of origin)? • Official determination of fire cause(s)? • Is the fire being investigated? • Agencies involved in fire investigation? • Are charges being brought against responsible parties? • *See chapter 23 for information on arson and incendiary fires.*
Sidebar Story Ideas	• Bulk storage facility fires • Chemical plant safety • Commodity flow studies • Community response plans • DOT hazardous materials classification system

Fig. 12–5. Media coverage job aid (continued)

	• Evacuation versus sheltering in place • Gas grill safety • Hazardous materials • Hazardous materials teams • Hazardous materials training and certification • Highway detour systems • Living in a world of hazardous materials • Local emergency planning committees (LEPCs) • Refinery fires • Transportation emergencies
Resources	• CHEMTREC www.chemtrec.com • Department of Transportation (DOT) www.dot.gov • Environmental Protection Agency (EPA) www.epa.gov • Federal Emergency Management Agency (FEMA) www.fema.gov • National Fire Protection Association (NFPA) www.nfpa.org • National Highway Transportation Safety Administration (NHTSA) www.nhtsa.gov • National Transportation Safety Board (NTSB) www.ntsb.gov • Occupational Safety and Health Administration (OSHA) www.osha.gov • United States Fire Administration (USFA) www.usfa.fema.gov

Fig. 12–5. Media coverage job aid (continued)

Chapter Questions

1. Identify the typical categories of hazardous materials fires.
2. List the incident management priorities in managing hazardous materials fires.
3. Discuss several operational tactics related to hazardous materials fires.
4. Identify the safety concerns related to hazardous materials fires.
5. Identify the emergency service and support organizations and agencies that are dispatched to hazardous materials fires.
6. Discuss the resource requirements, in terms of personnel, apparatus, and equipment, associated with hazardous materials fires.
7. Relate and explain the essential story elements of news coverage of hazardous materials fires.

Chapter 13
Outside Fires

Chapter Objectives

- Identify the typical categories of outside fires.
- Identify the incident management priorities for outside fires.
- Discuss the operational tactics utilized at outside fires.
- Discuss the safety considerations associated with outside fires.
- Identify the emergency service agencies that respond to outside fires.
- Identify the resource requirements associated with outside fires.
- Discuss the story elements associated with media coverage of outside fires.

The National Fire Protection Association (NFPA) estimates that there were 686,000 "outside and other fires" in the U.S. in 2011, resulting in an estimated property damage cost of $616 million. These incidents resulted in a reported 65 civilian deaths and 675 civilian injuries.[1] While fires considered outside fires routinely occur, it is not unusual that some of these fires result from their proximity to buildings, vehicles, or equipment involved in fire or correspondingly present exposure hazards to such properties.

Typical Incidents

As shown in figure 13–1, fires that are considered "outside fires" fall within a fairly broad continuum of incidents ranging from a small mulch or grass fire adjacent to an office complex parking lot to a sizable wildland fire capable of consuming thousands of acres of land and anything contained on that land. The common theme that runs through each of these incidents is that they take place outside of a building or structure. The property on which these incidents occur can be used for various purposes, as in the case of a tire fire occurring in an outside storage area at a tire reclamation facility or a fire in a junkyard involving the vehicles housed within the property. Fires that occur in dumps or landfills likewise fall into this fire category, as do those fires involving docks or piers.

- Brush fire
- Controlled burning (authorized)
- Dock or pier fire
- Dump or landfill fire
- Dumpster/trash compactor fire
- Fireworks explosion
- Forest fire
- Grass fire
- Junkyard fire
- Outbuilding fire
- Outside tire fire
- Prescribed fire
- Unauthorized burning
- Wildland fire

Fig. 13–1. Typical incidents: outside fires

Fires often occur in trash dumpsters as well as in trash compactors. These fires can present exposure hazards if a dumpster is located in close proximity to a building or other property. It is common that trash compactors are installed to permit the placement of materials into the compactor from the inside of a building and transfer the resulting compacted material into a large dumpster or container on the outside of the building. Based on their design, trash compactors can present exposure hazards in the event of a fire involving their machinery or contents. Outbuildings, such as sheds, also fall into the outdoor fire classification. Given that they often exist to store items such as pesticides and other chemicals that property owners prefer not to store within another building, the potential inherent hazards resulting from the hazardous materials that they may house are often disproportionate to their small footprint on the property.

A sizable number of outside fires are considered wildland fires. These fires can be further categorized based on the nature of the vegetation that fuels such fires, ranging from grass fires to brush fires to forest fires (fig. 13–2). These categories of wildland fires have both common and unique characteristics in terms of the firefighting challenges that they present. Grass fires, for example, typically take place in locations that are more accessible in terms of available water supply and access for fire apparatus. In contrast, forest fires often occur in remote areas that are difficult to access and may have a limited water supply.

Fig. 13–2. Wildland fire

There will be times when the lighting of a fire is sanctioned by the authority having jurisdiction (AHJ), as in the case of an authorized controlled burning or a prescribed fire that is properly planned and conducted in accordance with a prudent and approved wildland fire management program. Human nature being what it is, there will also be times when individuals will light fires without receiving prior approval or in violation of community fire regulations. The causes of outside fires fall into the categories of natural, accidental, or intentional. Lightning is the cause of many outside fires each year. Likewise, a significant number of outdoor fires occur each year as a result of careless smoking or not properly attending to campfires and other permitted fires. Fireworks accidents and explosions can ignite outside fires, as well as structure fires. Unfortunately, some outside fires are started intentionally by individuals with a range of motives. In all cases where an incendiary fire is suspected, a fire investigation should be initiated.

Incident Priorities

As with all incidents, the priorities in the management of an outside fire, in priority order, are (1) life safety, (2) incident stabilization, and (3) property conservation.

Operational Overview

As with other types of fire, all decisions and actions involved in fighting an outside fire will be based on the priority of life safety. An essential element of all related firefighting operations will be to ensure that firefighters are not given assignments or permitted to operate from locations that would compromise their life safety. Examples of this would be the application of extinguishing agents from safe distances in instances such as dumpster, trash compactor, junkyard, or outside tire fires. To do otherwise would present an unacceptable risk to firefighting personnel, particularly in those situations where the incident presents no danger to the life safety of the public. The potential presence of hazardous materials in any outside fire must always be considered in the selection of operational tactics.

A strategic goal with all outside fires is to prevent the spread of fire to exposures, whether from a dumpster to an adjacent building, vehicle, equipment, or infrastructure such as power lines, or from a small campfire in a national park rapidly evolving into a massive forest fire. A contributing factor in a growing number of wildland fires responsible for damaging not only the land and its vegetation, but also the homes and other structures that have been built upon it, is what has come to be called the urban/wildland interface. This interface area, which is highly susceptible to fire spread and damage, has resulted from aggressive residential development that extends into wooded lands, including forests. Individuals building these homes typically do so in the interest of the rustic setting that their embedded locations offer. Many likewise seek to fully achieve the rustic appearance by using wood building materials, including roofing shingles, which aesthetically blend into the rustic surroundings of the residence. While these homes routinely result in beautiful and peaceful places to live, that can all change quickly and without warning upon the occurrence of a fire that rapidly spreads through the wildlands and consumes the houses that have been so perfectly implanted within them. In addition to the building and finishing materials that routinely create fire hazards associated with the construction of residences in the urban/wildland interface is the fact that either in their original design and construction, or their subsequent property maintenance, appropriate provisions are not incorporated to provide protection barriers with respect to the spread of fire.

While the general principles of firefighting that apply to other types of firefighting will likewise apply in situations involving outside fires, the nature and specifics of a particular outside fire will determine the appropriate approach that should be taken in the interest of resolving the incident. Based on the size-up of the incident, the incident commander will, in accordance with established incident priorities, formulate and implement a plan to resolve the incident in a manner that results in effective, efficient, and safe extinguishment of the fire. The required tactics will usually involve containing or confining the fire, extinguishing it, and performing appropriate overhaul in the interest of preventing a reignition or rekindle of the fire. In so doing, a water supply must be established and water and other appropriate extinguishing agents applied. Firefighting activities will also emphasize exposure protection, especially in cases where an outside fire has the potential of spreading to structures. In the interest of ensuring the life safety of the public, perimeters must be established and enforced, vehicular traffic must be controlled, and evacuations may be required. Actions to provide for the life safety of emergency responders include the use of an incident management system and a personnel accountability system and the provision for rehabilitation. As always, appropriate personal protective equipment, including respiratory protection, should be worn, as in the case of firefighters wearing their structural turnout gear and self-contained breathing apparatus (SCBA) when fighting a dumpster, trash compactor, junkyard, or similar outside fire.

While many outside fires are fairly routine and within the capabilities of firefighters who are trained in structural firefighting and routinely operate in that environment, the world of wildland firefighting is significantly different in a number of ways, including appropriate personal protective equipment and firefighting tactics. Fire departments whose response area includes wildlands have recognized the importance of providing supplemental wildland firefighting training to their personnel. This is a prudent decision in that fallen firefighter trends have shown that deploying experienced structural firefighters into the wildland firefighting environment without appropriate supplemental training can yield tragic results. This chapter focuses on providing a general understanding of wildland firefighting. Media professionals whose assignments are likely to include covering wildland fires will benefit from the resources listed in figure 13–5.

The incident management of a wildland fire begins with a thorough size-up of the incident in terms of what has happened, what is happening now, and what is likely to happen. Factors considered in this evaluation include the fuel load in terms of vegetation (grass, brush, forest) and its fire characteristics, the size of the fire and area involved, potential exposures including structures and other property, fire intensity, rate of fire spread, and current and predicted weather conditions. Each of these size-up factors is essential in terms of selecting and implementing appropriate tactics as well as in ensuring firefighter safety. These factors are also interrelated in that the nature of the vegetation will play a significant role in determining both the fire intensity and rate of fire spread. Weather conditions can be contributing factors to the starting of a fire, the spread of the fire, and to the success of firefighters in controlling and extinguishing the fire. Relevant weather conditions include air temperature, humidity, wind direction, and wind speed. Both current and forecasted weather must be constantly monitored, given that expected and unexpected weather changes, including storms and lightning, may dictate the need for adjustment of operational tactics as well as presenting life safety issues for firefighters.

While the mental images that many have of wildland firefighting are those of dramatic air drops of water and other fire extinguishing agents and of mobile attacks by specially designed brush trucks with pump-and-roll capabilities, the vast majority of wildland firefighting involves the highly labor-intensive work of wildland firefighters walking in, being driven in, or being air dropped into a wildland fire setting. While the air and mobile attacks can have a significant impact in containing a fire, it is the strenuous work and heroic efforts of the firefighters on the ground that results in the extinguishment of the fire. These firefighters often operate in remote areas with limited resources, including water supply, and thus may often have to primarily or exclusively use hand tools to extinguish and overhaul a fire (fig. 13–3). In so doing they will seek to work from the area where the fire originated that is now burned to the area where the fire has the greatest rate of spread, called the head, as it advances toward the unburned area. In so doing, firefighters will attempt to stop the fire through the use of fire control lines—natural boundaries or those that they have created in the interest of stopping the spread of fire and enabling its extinguishment. At times, authorized and appropriately trained firefighters will use backburning, whereby a section of the wildlands is burned prior to the fire's arrival in the interest of creating a fire break with no remaining fuel to burn and support the fire.

Fig. 13–3. Wildland firefighter

Scene assessment and management are essential elements of the successful management of a wildland fire. A thorough size-up must be continuously conducted throughout the incident with an ever-present evaluation of current and predicted weather conditions that could favorably or unfavorably affect firefighting operations, success, and firefighter safety. Comprehensive resource management and logistical support are essential at many wildland fires given that they can extend over days, weeks, or even months. Staging of incoming resources is a must in the successful management of these incidents.

Safety Considerations

Outside fires present many issues that can compromise firefighter safety. This is particularly true in the case of wildland fires. All firefighting activities related to outside fires must be conducted in a manner that ensures firefighters are properly trained and equipped to effectively, efficiently, and safely complete the assignments that they are given. This begins with having the necessary and appropriate training and recognizing that wildland firefighting and structural firefighting require different approaches, tactics, and tools of the trade. Whereas structural firefighters rely on the protective environment provided by their bulky structural turnout gear, firefighters involved in wildland firefighting need lightweight protective clothing that, while providing thermal protection, affords them the necessary mobility and endurance to complete their tasks without becoming fatigued given the long durations and physical demands of their firefighting activities. Personal protective equipment for wildland firefighters typically includes fire retardant pants, coats, shirts, jumpsuits, or overalls; protective hoods; gloves; eye protection; helmets; and a personal fire shelter designed to be wrapped around the firefighter as a last resort if he or she were to become trapped or be in danger of being overrun by a fire. The fact that many wildland fires occur in remote areas that are not accessible by motorized vehicles, thus requiring that firefighters hike in and out or be air dropped in, likewise affords the potential for fatigue or injuries. Maintaining personnel accountability, crew integrity, and providing necessary rehabilitation are vitally important at these incidents. The air temperature and humidity have a direct correlation with the environmental demands of a wildland incident that accompany the labor-intensive physical demands of wildland firefighting.

The fuel load, fire intensity, and rate of fire spread contribute to safety concerns that must be addressed by those commanding an incident and those supervising crews operating at the incident. A number of weather conditions—air temperature, humidity, wind direction, and wind speed—contribute to the intensity and rate of spread of a wildland fire. While the current weather conditions are extremely important to monitor, it is equally if not more important to monitor and incorporate changing weather conditions into operational decisions. It is essential that communication be maintained will all crew members throughout the incident in the interest of ensuring their safety and providing needed assistance should they become disoriented, lost, or trapped. Communication problems can significantly compromise life safety in instances where personnel accountability cannot be verified. All tactical operations on the incident scene, which in some cases can span many acres and miles, must be coordinated given the integrated nature of these operations. Freelancing cannot be tolerated.

The nature and dynamics of wildland fires produce additional safety challenges. These include firefighters being overrun by fast-moving fires, such as crown fires that ignite and rapidly spread in the tops of trees independent of the lower fire spread. Additional safety concerns are fires that jump natural or artificial barriers, falling trees weakened by the fire, and power lines compromised by the fire. As with any fire, the potential for smoke inhalation exists. Although air drops of extinguishing agents serve an essential purpose at many wildland incidents, these air drops can compromise firefighter safety if personnel are not clear of the drop zone and are thus hit by the extinguishing agent or if the drop causes trees or branches to break and fall on firefighters.

While the appointment of an ISO, often supported by assistants, is obviously important in these incidents as in all others, wildland fires are incidents that illustrate the importance of crew leaders taking responsibility for the safety of their crews and of each firefighter looking out for his or her own safety as well as the safety of other crew members. An essential element of ensuring firefighter safety at these incidents is always to identify and ensure the viability of areas of refuge, safety zones, and escape routes.

Resource Requirements

The typical response to an outdoor fire will be the fire department. Based on its nature and size, additional resources may be sent on the initial alarm or as subsequent mutual aid. These additional agencies will typically include forest fire agencies, law enforcement, and emergency medical services. While the necessary apparatus and equipment will be based on the nature of the incident, the apparatus typically dispatched to wildland fires will include brush trucks, wildland engines, mobile water supply apparatus, and emergency medical units. The apparatus needs of wildland fires may also include fixed-wing aircraft, helicopters, specialized apparatus, support apparatus, and land moving equipment. It is customary for crews to be assembled and deployed from the staging area as strike teams or task forces. The typical tools used by firefighters at outdoor fires include chain saws, deck guns, extinguishing agents, hand tools, handlines, portable equipment, specialized wildland firefighting hand tools (such as tools that offer the functionality to cut trees and brush as well as to dig or trench the ground), supply hose, traffic control equipment, and water tank extinguishers.

Media Coverage of Outside Fires

While the impact of many smaller, isolated outside fires will be relatively minor and of little interest to the public and the news media, some outdoor fires, such as larger wildland fires, may be of great interest to a number of different stakeholder groups, including the news media. Members of the public may fall into one or more stakeholder categories based on their specific need for or interest in learning more about a wildland fire. Those whose commute to work or school may be affected by the road closures and detours resulting from the fire and would obviously be interested in receiving up-to-date information in a timely and accurate manner. This is an example of how a media organization's news and traffic reporters can collaborate in the interest of providing the accurate, comprehensive, professional, and timely media coverage that all stakeholders expect and deserve. Likewise, those individuals planning to travel in the impacted area, whether for a family vacation or on business, would be very interested in fully understanding the evolving situation and impact of the fire in the interest of adjusting their travel plans as necessary. Taking the family on a long-awaited vacation to a national park only to discover upon arrival that it was closed due to the fire would likely rival the disappointment of the Griswold family in the movie *Summer Vacation* upon their arrival at Wallyworld, only to find that the park was closed for maintenance.

Those individuals whose homes are in immediate, imminent, or potential danger depending on the further spread of the fire and the success of firefighters in stopping the fire's advancement are obviously those with the greatest interest in staying continuously informed regarding the fire in terms of what has happened, what is happening now, and what is likely to happen. Given the role that changing weather conditions, such as wind direction or speed, can play with respect to the fire's behavior and potential for fire spread, this is an appropriate opportunity for news reporters and meteorologists to collaborate in the news coverage and updates. Radio and television stations whose coverage area is involved will usually interrupt scheduled programming for breaking news reports and updates. When deemed appropriate based on the severity of the incident and its potential for harm to a community or region and its residents, they may opt to suspend regular programming in the interest of serving the impacted community through the provision of continuous news coverage.

In events like this, the news media, working collaboratively with fire and emergency service organizations, emergency management agencies, and elected and appointed officials, play an instrumental role in contributing to the life safety of the community by incorporating informational updates, alerts, and instructions into their news coverage. Providing accurate information to the public on when, where, and how to evacuate is essential from the standpoint of ensuring the life safety of the public and contributing to the effective, efficient, and safe management of the incident. This information should include information on available shelters to which individuals can evacuate, and their capabilities, including the ability to accept individuals with certain types of medical conditions and those who have pets that they desire to bring with them to a shelter. Information about detours and routes of travel for evacuations that do not interfere with the response of fire and emergency service resources can be provided. Essential elements of the news coverage of these incidents will correlate with the primary issues discussed throughout this chapter that contribute to the incident and must be anticipated and addressed in its successful management—fire conditions, current and projected weather conditions, progress in containing the fire, and anticipated fire behavior and spread.

In addition to the informative coverage that the public counts on the news media to provide during incidents such as major wildland fires, the fire department and the news media have an opportunity to provide proactive public information and education designed to help prevent such fires from occurring in the first place. This could involve the provision of information on the impact of careless smoking or campfires. Relevant cautions and messages on outdoor fire safety could be incorporated into weather forecasts on days when the aforementioned current or forecasted weather conditions have the potential of leading to the ignition and spread of a wildland fire. Additionally, sidebar stories that discuss how residents, particularly those living in an urban/wildland interface area, can take appropriate actions to make their properties and homes defensible in the event of a wildland fire can be run by the news media.

The job aids in figures 13–4 and 13–5 are designed to serve as resources for fire department and media personnel. Figure 13–4 provides an overview of incident management of outside fires, while figure 13–5 provides guidance with respect to media coverage of these incidents.

Incident Management: Outside Fires	
Typical Incidents	• Brush fire • Controlled burning (authorized) • Dock or pier fire • Dump or landfill fire • Dumpster/trash compactor fire • Fireworks explosion • Forest fire • Grass fire • Junkyard fire • Outbuilding fire • Outdoor tire fire • Prescribed fire • Unauthorized burning • Wildland fire
Incident Priorities	1. Life safety 2. Incident stabilization 3. Property conservation
Operational Tactics*	• Air attack, backburning, confinement, containment, evacuation, exposure protection, extinguishment, fire control lines, fire investigation, logistical support, mobile attack, overhaul, perimeter control, personnel accountability, rehabilitation, resource management, scene assessment, scene management, staging, traffic control, and water supply
Safety Considerations*	• Accessibility, air drops of extinguishing agents, areas of refuge, being overrun by fire, communication problems, crown fires, current and forecasted weather conditions (air temperature, humidity, wind direction, wind speed), escape routes, falling trees, fire intensity, fire jumping natural or manmade barriers, firefighter fatigue, firefighters becoming disoriented, lost, or trapped, freelancing, fuel load, hazardous materials, injuries, involvement of structures, limited water supply, long-duration incidents, personnel accountability, physical demands on personnel, power lines, rate of fire spread, rehabilitation, rekindle, remote areas, safety zones, smoke inhalation, spot fires, and unexpected weather changes
Responding Agencies*	• Fire department, forest fire agencies, law enforcement, emergency medical services

Fig. 13–4. Incident management job aid. **Note:* The nature of a particular incident will determine the appropriate operational tactics, safety considerations, responding agencies, and resource requirements, with wildland firefighting being significantly different from other outside firefighting.

Resource Requirements	Personnel* • Fire department, wildland firefighters, law enforcement, emergency medical services, specialized equipment operators, and logistical support personnel. Crews will often be assembled as strike teams or task forces. Personal protective equipment* • Appropriate personal protective equipment will be based on the nature of the incident, with normal structural firefighting turnout gear, including self-contained breathing apparatus (SCBA) being appropriate in incidents such as dumpster, junkyard, or outdoor tire fires. PPE for wildland firefighting typically includes fire retardant pants, coats, shirts, jumpsuits, or overalls; protective hoods; gloves; eye protection; helmets; and a personal fire shelter. Apparatus* • Brush trucks, wildland engines, mobile water supply apparatus, fixed-wing aircraft, helicopters, specialized apparatus, land moving equipment, support apparatus, and emergency medical units Equipment* • Chain saws, deck guns, extinguishing agents, hand tools, handlines, portable equipment, specialized wildland firefighting hand tools, supply hose, traffic control equipment, and water tank extinguishers

Fig. 13–4. Incident management job aid (continued)

Media Coverage: Outside Fires	
WHO?	• Person(s) reporting fire? • Responsible party identified? • Personal information on responsible party? • Activity of responsible person at time fire started? • Person(s) of interest? • Number of victims? • Number of and extent of civilian injuries? • Number of and cause of civilian fatalities? • Number of and extent of firefighter injuries? • Number of and cause of firefighter fatalities? • Number of missing persons? • Personal information on injured individuals? • Personal information on fatalities? • Personal information on missing persons?

Fig. 13–5. Media coverage job aid

	• *Have appropriate notifications been made?* • *Has information on victims been received from a credible and authorized source?* • *Has information been verified for accuracy?* • How and to which medical facilities were victims transported? • Medical condition of victims (initial and present)? • *See chapter 21 for incidents involving civilian injuries or fatalities.* • *See chapter 22 for incidents involving firefighter injuries or fatalities.*
WHAT?	• Type of fire? • Wildland fire? • Nature of vegetation (grass, brush, forest)? • Area involved (approximate square feet or acres)? • Area threatened or exposed? • Number of involved buildings? • Number of threatened or exposed buildings? • Number of involved vehicles? • Other involved property? • Other threatened or exposed property? • Damage to buildings? • Damage to vehicles? • Other damage? • Threatened wildlife? • Domestic animal casualties? • Wildlife casualties? • Responding agencies? • Incident priorities? • Operational tactics? • Safety considerations? • Incident management challenges? • Resource requirements (number and types of personnel, apparatus, and equipment)? • Assisting agencies? • Cooperating agencies? • Availability and adequacy of water supply? • Hazardous materials involvement? • Resulting health or environmental concerns? • Evacuation or shelter in place? • Evacuation routes? • Road closures or lane restrictions? • Traffic detours?

Fig. 13–5. Media coverage job aid (continued)

WHEN?	• Day of week? • Time of day? • Day/night? • Duration of incident?
WHERE?	• Incident location? • Nearby recognized landmarks (roads, buildings, etc.)? • Alternate location references (route number vs. street name)? • Municipality? • Proximity to past incidents? • Urban/wildland interface area? • State property? • Federal property? • National park? • National forest? • Historical site? • Property use?
WHY?	• Factors contributing to fire start? • Weather conditions (air temperature, humidity, wind direction, wind speed)? • Lightning? • Fire danger rating? • Fuel load? • Unusual vegetation growth? • Drought conditions? • Fuel moisture? • Weather forecast (air temperature, humidity, wind direction, wind speed)? • Prescribed fire? • Human factors? • Careless smoking? • Careless cooking or campfire? • Activities contributing to fire? • Official determination of cause(s)? • Fire under investigation? • Agency handling fire investigation? • Charges brought against responsible parties? • *See chapter 23 for information on arson and incendiary fires.*

Fig. 13–5. Media coverage job aid (continued)

HOW?	• Factors contributing to fire spread? • Fire intensity? • Rate of fire spread? • Land topography/grading?
Sidebar Story Ideas	• Building and maintaining a defensible residence • Campfire safety • Careless smoking • Evacuation • Evacuation routes • Fire protection in national parks and forests • Fire safety during drought conditions • Forest fires • Incendiary fires • Lightning as a fire cause • Living in the urban/wildland interface • Outdoor fire safety • Sheltering in place • Stories on major wildland fires • United States Forest Service • Urban/wildland interface • Wildland firefighters • Wildland firefighting
Resources	• National Fire Protection Association (NFPA) www.nfpa.org • United States Fire Administration (USFA) www.usfa.fema.gov • United States Forest Service (USFS) www.fs.fed.us

Fig. 13–5. Media coverage job aid (continued)

Chapter Questions

1. Identify the typical categories of outside fires.
2. List the incident management priorities in managing outside fires.
3. Discuss several operational tactics related to outside fires.
4. Identify the safety concerns related to outside fires.
5. Identify the emergency service and support organizations and agencies that are dispatched to outside fires.
6. Discuss the resource requirements, in terms of personnel, apparatus, and equipment, associated with outside fires.
7. Relate and explain the essential story elements of news coverage of outside fires.

Notes

1 National Fire Protection Association. 2011. *National Fire Protection Association Estimates*. Quincy, MA: National Fire Protection Association.

Chapter 14

Vehicle Fires

Chapter Objectives

- Identify the typical categories of vehicle fires.
- Identify the incident management priorities for vehicle fires.
- Discuss the operational tactics utilized at vehicle fires.
- Discuss the safety considerations associated with vehicle fires.
- Identify the emergency service agencies that respond to vehicle fires.
- Identify the resource requirements associated with vehicle fires.
- Discuss the story elements associated with media coverage of vehicle fires.

While many of the calls that a contemporary fire department is dispatched to could be classified as vehicle fires, the nature of these incidents and the corresponding incident management challenges that they may present vary greatly based on the specific vehicle(s) involved and related situational aspects of the incident. Each of the various makes, models, and types of vehicles presents its own unique challenges to firefighters. The National Fire Protection Association (NFPA) estimates that in 2011 there were 219,000 vehicle fires in the United States and that these fires resulted in 1,190 civilian injuries, 300 civilian deaths, and a direct dollar property loss of $1.4 billion.[1]

While the focus of this chapter is on vehicles that operate on our nation's highways, other transportation conveyances, including those using air, rail, and water, likewise present the potential for vehicle fires. The presence or involvement of hazardous materials when responding to a vehicle fire will often result in the decision to operate in a different operational mode and utilize different tactics than would be the case were the incident to not involve such materials and their accompanying hazards. Further consideration of hazardous materials fires and incidents can be found in chapters 12 and 15, respectively.

Typical Incidents

The vehicles involved in incidents to which fire departments respond come in all types and sizes based on their purpose, as listed in figure 14–1. Many of these vehicles are designed for the transport of passengers in either small or large groups. Passenger vehicles include automobiles, pickup trucks, sport utility vehicles, and vans that provide transportation for a limited number of passengers. Passengers also travel in buses, trains, planes, boats, and ships, which present the potential of high life hazards, as in the case of a fire on a school bus transporting students or a fire on a cruise ship. Many vehicles are designed for the transport of various types of cargo and commodities, as in the case of the various types of trucks that range from small cargo vans and box trucks to the tractor trailers that travel our nation's highways on a daily basis. Additionally, there are numerous types of specialized vehicles, ranging from recreational vehicles used by a vacationing family to the array of impressive machines that fall within the heavy construction equipment category.

While the diversity of contemporary transportation vehicles is obvious, they can be categorized according to some common characteristics. One such commonality is that each vehicle has a purpose—either transporting people or cargo—and both present incident management challenges. Another common characteristic is how the vehicle is powered. Its energy or power source will likewise influence many of the decisions of the incident commander at a vehicle fire. Additional considerations in incident management include the location and cause of the incident. The vehicle may have been parked when the fire occurred or in transit. Oftentimes, the fire may have been triggered as a result of a vehicle accident, perhaps resulting in the vehicle leaking its fuel or hazardous materials cargo.

- Aircraft fire
- Bus fire
- Heavy construction equipment fire
- Passenger vehicle fire (automobile, pickup truck, sport utility vehicle, van)
- Railcar fire
- Recreational vehicle fire
- Shipboard fire
- Tank truck fire
- Truck fire
- Truck and semi-trailer fire
- Water vehicle fire
- Alternate fuel vehicles
- Hybrid vehicles
- Fires while in transit
- Fires while parked
- Fires resulting from accidents

Fig. 14–1. Typical incidents: vehicle fires

The presence and/or involvement of hazardous materials, including the fuel powering the vehicle, and its cargo or contents, are crucial dimensions of these incidents. The new technologies that are revolutionizing contemporary vehicles, including hybrid and alternate fuel vehicles, likewise can present formidable challenges to firefighting strategy and tactics. While hybrid vehicles combine the two traditional technologies of a gas combustion engine and an electrical motor, alternative fuel vehicles derive their power from a growing number of fuels, including compressed natural gas (CNG), liquid propane gas (LPG), liquid natural gas (LNG), and hydrogen (H) (fig. 14–2). The causes of vehicle fires include accidents, batteries and electrical systems, exhaust systems, mechanical failures and malfunctions, and fuels, with some fires likewise being intentionally set. Human factors also contribute to vehicle fires, as in the case where a vehicle operator parks a vehicle with an extremely hot catalytic converter over ground cover that can easily ignite and subsequently involve the vehicle.

Fig. 14–2. Alternate fuel vehicle

Incident Priorities

As with all incidents, the priorities in the management of a vehicle fire, in priority order, are (1) life safety, (2) incident stabilization, and (3) property conservation.

Operational Overview

The successful incident management of a vehicle fire, as with all incidents, begins with a thorough size-up of the incident that reveals what has happened, what is happening now, and what is likely to happen. It might be found that subsequent to a two-vehicle automobile accident, one of the vehicles caught on fire and that while the passengers of the other vehicle were able to self-extricate themselves from their car, one person is trapped in the vehicle that has a fire in the engine compartment that will likely spread to the passenger compartment in a fairly short period of time. Likewise, a vehicle fire involving a tractor trailer may presently be confined to the engine, cab, or a wheel/brake area but has the potential of spreading to the hazardous cargo that the vehicle is transporting. These examples illustrate the importance of assessing the presence of passengers and the nature of the vehicle's cargo or contents. The location of the accident will likewise be a crucial factor in incident management decision making, an example being the different life safety issues presented when operating at a vehicle fire in a parking lot rather than on a high-speed divided highway (fig. 14–3).

Fig. 14–3. Vehicle fire on a highway. (Courtesy of David Paul Brown, Montgomery County Department of Public Safety.)

Ensuring the life safety of firefighting personnel begins with staging all personnel and apparatus uphill and upwind from the fire. This is a must from the standpoint of not becoming part of the incident in the event of the release of hazardous materials either from an automobile gasoline tank, the diesel fuel saddle tanks on a commercial truck, or the release of the hazardous materials that the vehicle was hauling. As discussed in chapter 12, the presence of hazardous materials will dictate an appropriate operational mode, strategic goals, and tactics. Likewise, the timely identification of the vehicle's power source is crucial in this age of hybrid and alternate fuel vehicles, given that each of these technologies can present hazards to firefighters that are both unique and significant.

Effective, efficient, and safe incident management of vehicle fires begins with providing a safe, protected area for firefighters to work, making scene assessment and management critically important. Control zones and perimeter controls should be established to provide for crowd and traffic control. It is essential that necessary actions be taken to ensure the stabilization of the vehicle so it does not move, shift, or roll from its present location, injuring firefighters or other people. The presence of hazardous materials and their associated properties must be identified in the interest of determining an appropriate operational mode and implementing appropriate tactics such as air monitoring, hazmat control, containment, spill control, runoff control, or vapor suppression. In some cases providing for the life safety of individuals in proximity to the incident may require sheltering in place or conducting evacuations.

As with many incidents involving fire, exposure protection and utility control may be necessary. The extinguishment of a vehicle fire will involve the use of appropriate extinguishing agents, fire attack, confinement, and extinguishment, all in accordance with established departmental policies and procedures. Water supply, as always, will be important, particularly in the case of an incident involving a large quantity of hazardous materials burning and those in unhydranted areas. The nature of many vehicle fires, as well as the construction, design, and materials in modern vehicles, often requires overhaul in the interest of ensuring that the fire is fully extinguished. Likewise, salvage may be enacted in the interest of property conservation of a vehicle's contents or cargo. It will not be uncommon for firefighters to have to engage in forcible entry to gain access to the area of the vehicle that is burning, whether an engine compartment, passenger compartment or cab, trunk, or cargo area. As always, life safety comes first, thus requiring the tracking of personnel throughout an incident and an emphasis on locating and rescuing victims early in the incident.

Safety Considerations

The size and challenges of vehicle fires, in terms of both incident management and life safety of responders and the public, can potentially span as diverse a continuum as do the types of vehicles. The location where the vehicle is found provides an important starting point in ensuring life safety. If the involved vehicle is in close proximity to moving traffic, traffic control must be a priority before placing emergency responders in harm's way. Each and every year firefighters are injured or killed while working on highways. The magnitude of this critical issue can significantly increase when motorists take their focus off the road and other traffic to get a good look at the vehicle fire. Fire department personnel should never operate in traffic lanes that have not been property controlled to ensure their safety. In addition to staging apparatus and personnel a safe distance uphill and upwind from the fire, apparatus positioning can incorporate the use of these large vehicles to block the roadway and provide protection for personnel. Weather conditions and decreased visibility resulting from the smoke generated by the fire are likewise concerns when protecting emergency responders operating on a highway. The establishment and enforcement of control zones is essential, given that vehicle fires are known to draw both attention and spectators.

Proper apparatus positioning must be coupled with safe and effective crew positioning in that personnel should approach the vehicle from a safe position, including not standing in front of energy-absorbing bumpers that can injure firefighters when exposed to extreme heat, and only approaching the vehicle after it has been properly stabilized. There are many other hazards associated with today's vehicles, such as the potential for firefighter injuries from vehicle materials, air bags that can deploy, vehicle struts that can rupture, magnesium components, and water-reactive materials. Alternate fuel and hybrid vehicles yield additional life safety issues that must be understood and factored into incident management decisions.

As always, the presence or involvement of hazardous materials presents safety issues in terms of combustible or flammable materials, producing a toxic atmosphere, the potential for personnel exposure, and the resulting need for decontamination. It is imperative that appropriate personal protective equipment and respiratory protection be worn by all personnel operating at vehicle fires, regardless of whether hazardous materials have been identified, given that the materials from which vehicles are constructed, their tires, and their fuel load dictate the need for this protection. In the case of fires involving cargo transport vehicles, the timely identification of the involved cargo through means such as placards displayed on the vehicle is crucial during the initial incident size-up. As discussed in the related hazardous materials chapters, the potential for a boiling liquid expanding vapor explosion (BLEVE) must be anticipated and where possible prevented in situations involving vehicle fires. Given the dynamic nature of any vehicle fire, continual size-up, proper supervision, the use of an ISO, and personnel accountability are prerequisites to ensuring life safety.

Resource Requirements

The typical response to a vehicle fire will include the fire department, emergency medical services, and law enforcement. Should the fire involve hazardous materials, a hazardous materials team may be dispatched to the incident scene. The nature of the incident and the involved vehicle(s) may necessitate requesting the response of aircraft or marine firefighting units. Often a towing service will be requested to remove the vehicle after the fire has been extinguished. As appropriate, fire investigators will be requested to conduct an investigation of the fire. The typical apparatus dispatched to a vehicle fire will include firefighting apparatus, with rescue apparatus, aerial apparatus, specialized apparatus, water supply apparatus, and emergency medical units being requested based on the particular resource needs of an incident. The typical tools used by emergency responders at these incidents include air monitoring equipment, cribbing, cutting tools, deck guns, extinguishing agents, fire extinguishers, foam generation and application equipment, hand tools, handlines, portable tools, power tools, prying tools, supply hose, traffic control equipment, thermal imaging cameras, and wheel chocks.

Media Coverage of Vehicle Fires

Vehicle fires, like all fires, tend to quickly attract interest and an audience based on their dynamic visual effects and the fact that they often occur in populated areas from which many spectators can be drawn, including those with cameras or smart phones in hand and committed to capturing some good images to share with their friends or send to a news organization interested in getting early images of an incident. As with all incidents, the volume of stories that a news organization is covering on a given day, as well as their size and scope, will be a prime factor in the determination of whether a particular vehicle fire is deemed a newsworthy story deserving of deploying the organization's scarce field reporters and photojournalists. While a car fire where no one was hurt will often not make the cut, had that same vehicle fire resulted in a death or been intentionally set to cover up a crime, a caravan of reporters from numerous media organizations would likely deem the story newsworthy and make the pilgrimage to cover it.

Likewise, if the vehicle involved in the fire was carrying toxic chemicals that threatened life safety and the local environment, or resulted in shutting down a major traffic corridor during rush hour, there would without question be significant interest on the part of the news media and their audience. Television coverage would likely incorporate both reporters on the ground and photojournalists in helicopters above the incident showing both the fire and the snarled highways in the area. News radio stations would likewise be interested in bringing their listeners timely reporting on the threat to the community and the traffic implications. This would be another example of where cooperative information dissemination by fire department and emergency management personnel and their news media counterparts could be vitally important in ensuring the life safety, health, and well-being of the community through issuing official situation updates, alerts, and instructions.

The job aids in figures 14–4 and 14–5 are designed to serve as resources for fire department and media personnel. Figure 14–4 provides an overview of incident management of vehicle fires, while figure 14–5 provides guidance with respect to media coverage of these incidents.

Incident Management: Vehicle Fires	
Typical Incidents	• Aircraft fire • Bus fire • Heavy construction equipment fire • Passenger vehicle fire (automobile, pickup truck, sport utility vehicle, van) • Railcar fire • Recreational vehicle fire • Shipboard fire • Tank truck fire • Truck fire • Truck and semi-trailer fire • Water vehicle fire • Alternate fuel vehicles • Hybrid vehicles • Fires while in transit • Fires while parked • Fires resulting from accidents
Incident Priorities	1. Life safety 2. Incident stabilization 3. Property conservation
Operational Tactics*	• *Stage personnel and apparatus a safe distance uphill and upwind from the fire.* • *The presence of hazardous materials should be determined as a prerequisite to selection of an appropriate operational mode and the development and implementation of strategic goals and tactical objectives.* • *Hybrid and alternate fuel vehicles may require different tactics.* • Air monitoring, confinement, containment, control zones, crowd control, decontamination, evacuation, exposure protection, extinguishment, fire attack, follow SOPs, forcible entry, hazmat control, identify hazardous materials and properties, overhaul, perimeter control, personnel accountability, runoff control, salvage, scene assessment, scene management, search, select appropriate operational mode, spill control, stabilization, staging, traffic control, utility control, vapor suppression, ventilation, victim rescue, and water supply • *See chapter 12 for information on hazardous materials fires.* • *See chapter 15 for information on hazardous materials incidents.*

Fig. 14–4. Incident management job aid. **Note:* The nature of a particular incident will determine the appropriate operational tactics, safety considerations, responding agencies, and resource requirements.

Safety Considerations*	• *Stage personnel and apparatus a safe distance uphill and upwind from the fire.* • *Personnel should not operate in traffic lanes that have not been controlled.* • *Personnel should approach vehicle from safe position and should not stand in front of energy-absorbing bumpers.* • *Hybrid and alternate fuel vehicles present specialized safety issues.* • Airbag deployment, alternate fuel vehicles, apparatus positioning, batteries, blocking roadway, boiling liquid expanding vapor explosion (BLEVE), burn injuries, cargo, combustible materials, contamination/decontamination, control zones, crew positioning, crowd control, downed wires, evacuation, explosion, exposure, flammable materials, fuel spills, hazardous materials, hot surfaces, hybrid vehicles, ignition sources, magnesium components, personnel accountability, smoke inhalation, tires, toxic atmosphere, utilities, vehicle contents, vehicle materials, vehicle stabilization, vehicle struts, vehicular traffic, victims, visibility, water-reactive materials, and weather
Responding Agencies*	• Fire department, emergency medical services, law enforcement, hazardous materials teams, airport fire units, marine fire units, fire investigators, and towing services
Resource Requirements	Personnel* • Fire department, emergency medical services, law enforcement, and specialized personnel as required (hazardous materials team, airport fire department, marine fire unit) Personal protective equipment* • Coat, pants, boots, helmet, gloves, eye protection, personal alert safety system (PASS), and self-contained breathing apparatus (SCBA) Apparatus* • Firefighting apparatus, rescue apparatus, aerial apparatus, specialized apparatus, water supply apparatus, and emergency medical units Equipment* • Air monitoring equipment, cribbing, cutting tools, deck guns, extinguishing agents, fire extinguishers, foam generation and application equipment, hand tools, handlines, hydraulic jacks, portable tools, power tools, prying tools, supply hose, traffic control equipment (traffic cones, traffic directional arrows, road flares), thermal imaging cameras, and wheel chocks

Fig. 14–4. Incident management job aid (continued)

Media Coverage: Vehicle Fires	
WHO?	• Vehicle operator? • Vehicle occupants? • Vehicle owner? • Number of vehicle passengers? • Number of victims? • Number of and extent of civilian injuries? • Number of and cause of civilian fatalities? • Number of and extent of firefighter injuries? • Number of and cause of firefighter fatalities? • Personal information on injured individuals? • Personal information on fatalities? • *Have appropriate notifications been made?* • *Has information on victims been received from a credible and authorized source?* • *Has information been verified for accuracy?* • How and to which medical facilities were victims transported? • Medical condition of victims (initial and present)? • *See chapter 21 for incidents involving civilian injuries or fatalities.* • *See chapter 22 for incidents involving firefighter injuries or fatalities.*
WHAT?	• Number of involved vehicles? • Types of vehicles? • Passenger vehicle? • Cargo vehicle? • Cargo being transported? • Presence of hazardous materials? • Involvement of hazardous materials? • Involved organizations (railroad, trucking company, airline, taxi company)? • Alternate fuel vehicle? • Hybrid vehicle? • Vehicle hazards? • Damage to vehicle? • Damage to contents or cargo? • Damage to other vehicles? • Damage to buildings? • Damage to bridges and other structures? • Damage to surroundings? • Responding agencies? • Incident priorities? • Operational tactics?

Fig. 14–5. Media coverage job aid

	• Safety considerations? • Incident management challenges? • Resource requirements (number and types of personnel, apparatus, and equipment)? • Resulting health or environmental concerns? • Evacuation or shelter in place? • Road closures or lane restrictions? • Traffic detours?
WHEN?	• Day of week? • Time of day? • Day/night?
WHERE?	• Incident location? • Nearby recognized landmarks (roads, buildings, etc.)? • Alternate location references (route number vs. street name)? • Proximity to past vehicle fires? • Vehicle in transit? • Vehicle parked?
WHY?	• Factors contributing to fire? • Human factors? • Equipment factors? • Mechanical failure or malfunction? • Vehicle fuel? • Vehicle cargo? • Weather-related factors? • Was fire the result of a vehicle accident?
HOW?	• How did the fire start? • Area where the fire started (area of origin)? • Area(s) to which fire spread? • Factors that contributed to fire spread? • Official determination of fire cause(s)? • Is the fire being investigated? • Agencies involved in fire investigation? • Are charges being brought against responsible parties? • *See chapter 23 for information on arson and incendiary fires.*
Sidebar Story Ideas	• Alternate fuel vehicles • Causes of vehicle fires • Dangers of vehicle fires • Highway safety when fighting vehicle fires • Hybrid vehicles • Intentionally set vehicle fires

Fig. 14–5. Media coverage job aid (continued)

	• Passenger vehicle fires • Preventing vehicle fires • Transport vehicle fires • Unintentional vehicle fires • Vehicle fire trends • Vehicle firefighting • Vehicle fires on highways
Resources	• Federal Aviation Administration (FAA) www.faa.gov • National Fire Protection Association (NFPA) www.nfpa.org • National Highway Traffic Safety Administration (NHTSA) www.nhtsa.gov • National Transportation Safety Board (NTSB) www.ntsb.gov • U.S. Coast Guard (USCG) www.uscg.mil • U.S. Department of Transportation (DOT) www.dot.gov • United States Fire Administration (USFA) www.usfa.fema.gov

Fig. 14–5. Media coverage job aid (continued)

Chapter Questions

1. Identify the typical categories of vehicle fires.
2. List the incident management priorities in managing vehicle fires.
3. Discuss several operational tactics related to vehicle fires.
4. Identify the safety concerns related to vehicle fires.
5. Identify the emergency service and support organizations and agencies that are dispatched to vehicle fires.
6. Discuss the resource requirements, in terms of personnel, apparatus, and equipment, associated with vehicle fires.
7. Relate and explain the essential story elements of news coverage of vehicle fires.

Notes

1 National Fire Protection Association. 2012. *National Fire Protection Association Estimates*. Quincy, MA: National Fire Protection Association.

Chapter 15

Hazardous Materials Incidents

Chapter Objectives

- Identify the typical categories of hazardous materials incidents.
- Identify the incident management priorities for hazardous materials incidents.
- Discuss the operational tactics utilized at hazardous materials incidents.
- Discuss the safety considerations associated with hazardous materials incidents.
- Identify the emergency service agencies that respond to hazardous materials incidents.
- Identify the resource requirements associated with hazardous materials incidents.
- Discuss the story elements associated with media coverage of hazardous materials incidents.

According to the National Fire Protection Association (NFPA), there were an estimated 397,000 hazardous materials incidents in the U.S. in 2010.[1] While these incidents comprise a fairly small number of responses and percentage of overall incidents, their potential to bring harm to a community certainly puts them on the radar screen of calls to which contemporary fire departments and other emergency response organizations must be prepared to respond and mitigate the problem in an effective, efficient, and safe manner. While hazardous materials incidents have the potential of resulting in property damage, the greater potential of their harm involves people and the environment.

Typical Incidents

While many of the incidents that a fire department responds to will be dispatched as hazardous materials incidents, commonly referred to as "hazmat" incidents, the possibility always exists that upon arrival on the incident scene and size-up of the situation, it may be discovered that an incident not dispatched as such actually involves hazardous materials. This could be the case with accidents, various types of fires (building, equipment, outside, or vehicle), and infrastructure emergencies.

Hazardous materials incidents may or may not involve the presence of fire. Chapter 12 is devoted to hazardous materials fires, while this chapter focuses on the general issues of hazardous materials incidents, including releases. Hazardous materials incidents can be categorized in a number of ways based on the material(s) involved, the nature of the release, and the location of the release, as listed in figure 15–1. While the contemporary world as we know it relies on hazardous materials for many of the necessities and luxuries of life, when not contained and controlled, these materials present the potential for harm and the need for an emergency response.

- Carbon monoxide incident
- Diesel fuel leak or spill
- Explosion
- Gas leak (natural gas or LPG)
- Gasoline leak or spill
- Hazardous materials leak
- Hazardous materials release
- Hazardous materials spill
- Overpressuration/rupture
- Radiation release
- Hazardous materials incident (fixed facility)
- Hazardous materials incident (rail tanker)
- Hazardous materials incident (road tanker)

Fig. 15–1. Typical incidents: hazardous materials incidents

These incidents occur at fixed facilities and range from a leaking propane cylinder on a home barbecue grill, to a much larger compressed gas cylinder used at an industrial plant, and in transportation incidents involving hazardous materials during transport of that same compressed gas by road or rail. The nature of a hazardous materials release can be classified as a product leak or spill or the rupture of the container protecting the hazardous materials as well as the environment. While the majority of hazmat incidents to which a fire department responds will be fairly routine, such as a gasoline or diesel fuel leak or spill, or a leaking propane cylinder, there will be times that fire and emergency services will face more challenging hazardous materials incidents based on the material(s) involved or the presence of fire or victims. Fire departments are also facing a growing number of responses to carbon monoxide incidents, both in residences and commercial properties.

The initial step in managing a hazardous materials incident is determining the involved material(s); their properties in terms of such essential aspects as their potential to bring harm based on flammability, explosion, and health effects; and the appropriate approach that should be taken to effectively, efficiently, and safely resolve the situation in accordance with the established incident priorities delineated below. The U.S. Department of Transportation has developed a comprehensive hazard classification system that divides all hazardous materials into one of the following nine classes.

- Class 1: Explosives
- Class 2: Gases
- Class 3: Flammable liquids
- Class 4: Flammable solids
- Class 5: Oxidizers
- Class 6: Poisons
- Class 7: Radioactive materials
- Class 8: Corrosives
- Class 9: Miscellaneous materials[2]

These classifications are displayed through an established system of placards and product labels. The NFPA 704 Marking System is also used to categorize hazardous materials in terms of their fire hazard, reactivity, health hazard, and specific hazards, such as being water-reactive or an oxidizer.[3] While multiple reference sources will routinely be utilized in researching involved hazardous materials before developing a plan of action to address a situation, a starting point in this research is usually the *Emergency Response Guidebook*, published and updated periodically by the U.S. Department of Transportation.[4] This is a valuable and inexpensive resource that, in addition to providing useful information to the initial first responders arriving at a hazardous materials incident, can also be extremely useful to the media professionals covering the incident. An essential source of specific information on any hazardous material is its material safety data sheet (MSDS) prepared by its manufacturer or distributor. Facility-specific hazardous materials information will also be available in community response plans and facility preplans (fig. 15–2).

Fig. 15–2. Community response plan

Incident Priorities

As with all incidents, the priorities in the management of a hazardous materials incident, in priority order, are (1) life safety, (2) incident stabilization, and (3) property conservation.

Operational Overview

The initial size-up that takes place in the management of all emergency incidents is of paramount importance with hazardous materials incidents, as is ongoing size-up throughout the incident in terms of monitoring both changing situational factors and progress being made toward resolving the situation. The incident commander will want to determine what has happened, what is happening, and what is likely to happen. The first two determinations are referred to as situational analysis; the last as forecasting.

The first step in this process will be to identify the hazardous material(s) involved, their properties, and the appropriate strategies and tactics that should be used in the management and resolution of the incident. Initial protective control zones will be established to ensure the health and well-being of emergency responders, the public, and the media. Three control zones—cold, warm, and hot—will be established. Staging and certain other activities will take place in the cold zone, with the warm zone serving as a transitional and support area between the cold and hot zones. Entry into the hot zone will be restricted to only necessary personnel with the appropriate training and equipment to conduct the necessary tactics (fig. 15–3). As appropriate, evacuations will be enacted or people will be sheltered in place based on a reasoned determination of the appropriate and practical approach to ensure their safety. The assistance of the news media may be requested in alerting and updating the public regarding the incident and the necessity to shelter in place or evacuate.

Fig. 15–3. Hazmat entry team entering hot zone

An appropriate operational approach will be determined based on a cost-benefit analysis driven by the incident priorities. This approach may involve a *defensive strategy* of containing the released material and protecting exposures, an *offensive strategy* of committing personnel to stop the actual release, or a *nonintervention strategy* in those cases where intervention would yield unacceptable results in terms of the health and safety of responders and the public and/or the environmental impact. This could be the case in a situation involving an extremely hazardous substance, a water-reactive material, or a material where the application of water would create a much more hazardous environmental impact. Situations where the hazardous materials release has resulted in downed, injured, or trapped individuals present additional challenges and the need for additional rescue and/or emergency medical resources.

There are several general principles of operating at any hazardous materials incident. All apparatus and equipment should be staged an appropriate distance from the incident, both uphill and upwind. Personnel approaching the incident should likewise approach from both uphill and upwind. An incident management system should be implemented, and the incident commander should utilize appropriate resources to track current and forecasted weather conditions, including wind direction and speed.

The nature of the incident, including the type of release and the involved material(s), will be considered in the selection of appropriate operational tactics, including absorption, containment, damming, diking, dispersal, neutralization, and runoff control. All personnel operating in the warm or hot zones must wear appropriate personal protective equipment that is compatible with the involved chemical(s), as well as don self-contained breathing apparatus (SCBA). Upon exiting the hot zone, all personnel and their equipment must be properly decontaminated. Throughout the incident, appropriate meters and instruments will be utilized to measure critical dimensions of the incident that have a bearing on responder and community safety, as well as successful incident management and resolution. As appropriate, fixed fire protection systems should be supported and utilities properly controlled.

Exposure protection is an important element of the successful management of many hazardous materials incidents. This may involve the application of large quantities of water, through fog streams to cool and protect adjacent tanks, vehicles, or buildings. The adequacy of the available water supply to deliver a sustained sufficient volume of water is an important consideration. In addition to establishing and enforcing the necessary control zones, it will be necessary to devote resources to both crowd and traffic control.

Safety Considerations

Life safety of emergency responders at a hazardous materials incident is addressed through the appointment of an ISO and the use of a personnel accountability system. Many hazardous materials incidents become long-duration incidents. The importance of providing proper rehabilitation for emergency responders operating on the scene is of the utmost importance. Medical surveillance should be conducted before and after all entries made into the hot zone. Throughout the incident the incident commander and the ISO will monitor changing conditions in the interest of ensuring responder safety. Only personnel with the appropriate training, certification, and equipment should perform certain tasks at a hazardous materials incident, including those assigned to the team(s) entering the hot zone. While certain defensive operations may be performed by personnel trained and certified at the hazardous materials operations level, many tasks, including those performed in the hot zone, must only be performed by hazardous materials technicians.

It bears repeating that all personnel, apparatus, and equipment should be staged uphill and upwind, and at a significant distance from the hazmat incident. In addition to ensuring that response personnel are not contaminated by the release, this is prudent from the standpoint of not introducing ignition sources in near proximity to the released materials. When assigned to perform particular tactics or tasks, personnel should likewise approach the incident from uphill and upwind. Necessary control zones should be established and enforced. Appropriate personal protective equipment, including respiratory protection, must always be worn when emergency responders are operating on the scene of a hazardous materials incident. The three routes through which individuals, including emergency responders, can be exposed are absorption, ingestion, and inhalation.

In situations involving a gas leak that is on fire, it is imperative that the source of the gas supply be shut off before an attempt is made to extinguish the fire. To do otherwise would greatly increase the hazards associated with the continued flow of gas and the potential of an ignition source such as a light switch or a vehicle's ignition system. The importance of not applying water to any water-reactive material, regardless of whether it is on fire, must be recognized.

Both the public and the news media should be denied entry into areas that present hazards or interfere with incident operations. A media area should be established and a PIO made available to brief the media throughout the incident.

Resource Requirements

The typical response to a hazardous materials incident will include the fire department, hazardous materials teams, emergency medical services, and law enforcement. Additional resources in the form of technical specialists, public works personnel, and hazardous materials cleanup contractors may be requested based on the nature and resource needs of the incident. Representatives of governmental agencies having jurisdiction, including regulatory agencies and health departments, may also respond to the incident. While the necessary apparatus and equipment will be based on the nature of the incident, the apparatus typically dispatched to hazardous materials incidents will include firefighting apparatus, rescue apparatus, water supply apparatus, specialized apparatus, and emergency medical units. The typical tools used by emergency responders at hazardous materials incidents include attack lines (handlines), air cascade systems, deck guns, extinguishing agents, foam generation and application equipment, gas detectors and other metering instruments, hand tools, portable equipment, positive pressure ventilation fans, smoke ejectors, and supply hose lines.

Media Coverage of Hazardous Materials Incidents

Hazardous materials incidents, by their very nature, can quickly attract interested individuals, as well as the news media that are there to gather and report the breaking news story to an interested public. Photojournalists on the ground or in news helicopters often capture vivid and captivating images of hazardous materials incidents that can be incorporated into the local and national news coverage. Given the inherent life safety hazards associated with hazardous materials incidents, it is imperative that a media area be established in a safe location within the cold zone and that a knowledgeable PIO be made available to brief media representatives on the initial and subsequent events. Hazardous materials incidents that result in civilian and/or emergency responder injuries and/or fatalities will usually receive extensive news coverage during and following the incident. Guidance on media coverage of incidents involving civilian injuries or fatalities is provided in chapter 21, with companion guidance on media coverage of emergency responder injuries and fatalities provided in chapter 22.

The public's interest in media coverage of a hazardous materials incident is likewise high if the incident is of long duration or interrupts transportation routes or systems, such as automobile, truck, bus, or rail routes. The news media serves an essential role in alerting the public to transportation detours, infrastructure interruptions, and information on community hazards and associated direction for the public to shelter in place or evacuate. The public is likewise interested in learning more about what happened and the potential effects that the incident may have on them in terms of health and safety, particularly if the incident occurred in near proximity to where they live, work, or travel. Coverage of the environmental impact of a hazardous materials release is also expected. As always, the public's expectation for news coverage of a hazardous materials incident includes accuracy, comprehensiveness, professionalism, and timeliness.

A unique expectation for these types of incidents is that coverage also be provided in an easy-to-understand way; this can obviously be difficult at times given the complex nature of hazardous materials chemistry and processes. This is where making available a knowledgeable PIO supported by appropriate technical experts is extremely important in getting the story right and reporting it correctly. Complete news gathering and reporting on some incidents may involve interviewing the incident commander, hazardous materials team personnel, transportation company personnel, and/or facility management or technical personnel. Given the high stakes that can be involved in terms of community health and safety, a hazardous materials incident is a breaking news story where it is imperative that reporting be accurate, timely, and credible. Through collaboration, emergency service representatives, in an effort coordinated by the PIO and their counterparts in the news media, can ensure that the information that is disseminated to the public, including instructions, alerts, and status updates, fully meets the expectations of all stakeholders.

The two job aids in figures 15–4 and 15–5 are designed to serve as resources for fire department and media personnel. Figure 15–4 provides an overview of incident management of hazardous materials incidents, while figure 15–5 provides guidance with respect to media coverage of these incidents.

Incident Management: Hazardous Materials Incidents	
Typical Incidents	• Carbon monoxide incident • Diesel fuel leak or spill • Explosion • Gas leak (natural gas or LPG) • Gasoline leak or spill • Hazardous materials leak • Hazardous materials release • Hazardous materials spill • Overpressurization/rupture • Radiation release • Hazardous materials incident (fixed facility) • Hazardous materials incident (rail tanker) • Hazardous materials incident (road tanker)
Incident Priorities	1. Life safety 2. Incident stabilization 3. Property conservation
Operational Tactics*	• Absorption, containment, cool exposed tanks, crowd control, damming, decontamination, diking, establish control zones (hot, warm, and cold), evacuation, exposure protection, identification of hazardous materials and properties, mechanical systems control, monitoring, neutralization, review preplan, rehabilitation, runoff control, scene assessment, scene management, search and rescue, selection of appropriate operational mode, shelter in place, staging, support fixed fire protection systems, traffic control, utility control, ventilation, and water supply
Safety Considerations*	• Approach from uphill, approach from upwind, do not extinguish gas leak fire until gas supply has been shut off, don appropriate personal protective equipment (PPE) including self-contained breathing apparatus (SCBA), identify hazardous material(s) involved and properties, contamination, control zones, explosion, exposure (absorption, ingestion, or inhalation), evacuation and/or shelter in place, fatalities, injuries, personnel accountability, respiratory protection, utilities, victims, and water-reactive materials
Responding Agencies*	• Fire department, emergency medical services, law enforcement, hazardous materials teams, technical specialists, public works, hazardous materials cleanup contractors, and representatives of governmental agencies having jurisdiction

Fig. 15–4. Incident management job aid. **Note:* The nature of a particular incident will determine the appropriate operational tactics, safety considerations, responding agencies, and resource requirements.

Resource Requirements	Personnel* • Fire department, emergency medical services, law enforcement, and hazardous materials teams. Hazardous materials personnel certified to appropriate level based on the incident (awareness, operations, technician, or specialist) Personal protective equipment* • Coat, pants, boots, helmet, gloves, eye protection, protective hood, specialized chemical protective equipment, personal alert safety system (PASS), and self-contained breathing apparatus (SCBA) Apparatus* • Firefighting apparatus, rescue apparatus, water supply apparatus, emergency medical units, and specialized apparatus Equipment* • Attack lines (handlines), air cascade system, deck guns, extinguishing agents, foam generation and application equipment, gas detectors and other monitoring instruments, hand tools, portable equipment, positive pressure ventilation fans, smoke ejectors, and supply hose

Fig. 15–4. Incident management job aid (continued)

	Media Coverage: Hazardous Materials Incidents
WHO? NOTE: Some hazardous materials incidents may also involve the rescue of injured or downed persons.	• Building owner? • Vehicle owner and/or operator? • Building occupants? • Number of victims? • Number of individuals displaced? • Number of and extent of civilian injuries? • Number of and cause of civilian fatalities? • Number of and extent of firefighter injuries? • Number of and cause of firefighter fatalities? • Number of missing or unaccounted for persons? • Personal information on injured individuals? • Personal information on fatalities? • Personal information on missing persons? • *Have appropriate notifications been made?* • *Has information on victims been received from a credible and authorized source?* • *Has information been verified for accuracy?* • How and to which medical facilities were victims transported?

Fig. 15–5. Media coverage job aid

	• Medical condition of victims (initial and present)? • *See chapter 21 for incidents involving civilian injuries or fatalities.* • *See chapter 22 for incidents involving firefighter injuries or fatalities.*
WHAT? NOTE: Many hazardous materials incidents also involve vehicles or buildings.	• Nature of incident? • Type of release? • Hazardous material(s) involved? • Properties of involved hazardous materials? • DOT hazard classification(s)? • Type of container? • Fixed facility? • Rail tanker? • Road tanker? • Container capacity? • Estimated release amount? • Physical state when released? • Related information? • Buildings involved? • Vehicles involved? • Other property involved? • Exposed buildings? • Other exposed property? • Extent of damage? • Estimated damage to building? • Estimated damage to contents? • Damage to other buildings or property? • Damage to vehicles? • Damage to vehicle cargo? • Are there injuries? • Are there downed, injured, or trapped individuals? • Are there fatalities? • Expected length of business interruption? • Responding agencies? • Resource requirements (number and types of personnel, apparatus, and equipment)? • Incident priorities? • Operational mode (offensive, defensive, or non-intervention)? • Operational tactics? • Safety considerations? • Incident management challenges? • Resulting health or environmental concerns? • Evacuation or shelter in place?

Fig. 15–5. Media coverage job aid (continued)

	• Location of evacuation shelters? • Number of individuals displaced? • Number of businesses displaced? • Road closures? • Expected duration of road closures? • Traffic detours? • *See chapter 9 for incidents involving accidents.* • *See chapter 10 for incidents involving building fires.* • *See chapter 14 for incidents involving vehicle fires.* • *See chapter 17 for incidents involving rescues.* • *See chapter 18 for incidents involving searches.*
WHEN?	• Day of week? • Time of day? • Day/night? • How long has incident been going on?
WHERE?	• Incident location? • Municipality or neighborhood? • Nearby landmarks (roads, buildings, etc.)? • Area affected? • Population density of affected area? • Wind direction and speed? • Impact of present and forecasted weather? • Area and extent of evacuation? • Public water supply (hydrants)? • Source of water supply? • Electric utility? • Gas utility? • Sewer utility? • Water utility?
WHY?	• Source of material release? • Factors that contributed to release? • Source of ignition? • Contributing human factors? • Weather-related factors? • Adequacy of water supply? • Problems with water supply?

Fig. 15–5. Media coverage job aid (continued)

HOW?	• How did the incident occur? • Official determination of incident cause(s)? • Is the incident being investigated? • Agencies involved in incident investigation? • Are charges being brought against responsible parties?
Sidebar Story Ideas	• Carbon monoxide incidents • Chemical plant safety • Commodity flow studies • Community response plans • Decontamination procedures • DOT hazardous materials classification system • Evacuation versus sheltering in place • Explosions • Hazardous materials • Hazardous materials leaks • Hazardous materials regulations • Hazardous materials releases • Hazardous materials response • Hazardous materials spills • Hazardous materials teams • Hazardous materials training and certification • Highway detour systems • Living in a world of hazardous materials • Local emergency planning committees (LEPCs) • Routes of chemical exposure • Transportation emergencies
Resources	• CHEMTREC www.chemtrec.com • Environmental Protection Agency (EPA) www.epa.gov • Federal Emergency Management Agency (FEMA) www.fema.gov • National Fire Protection Association (NFPA) www.nfpa.org • National Transportation Safety Board (NTSB) www.ntsb.gov • Occupational Safety and Health Administration (OSHA) www.osha.gov • United States Fire Administration (USFA) www.usfa.fema.gov

Fig. 15–5. Media coverage job aid (continued)

Chapter Questions

1. Identify the typical categories of hazardous materials incidents.
2. List the incident management priorities in managing hazardous materials incidents.
3. Discuss several operational tactics related to hazardous materials incidents.
4. Identify the safety concerns related to hazardous materials incidents.
5. Identify the emergency service and support organizations and agencies that are dispatched to hazardous materials incidents.
6. Discuss the resource requirements, in terms of personnel, apparatus, and equipment, associated with hazardous materials incidents.
7. Relate and explain the essential story elements of news coverage of hazardous materials incidents.

Notes

1 National Fire Protection Association. 2011. *National Fire Protection Association Estimates*. Quincy, MA: National Fire Protection Association.

2 U.S. Department of Transportation. 2012. *Department of Transportation Hazard Classification System*. Washington, DC: U.S. Department of Transportation.

3 National Fire Protection Association. 2012. *NFPA 704: Standard System for the Identification of the Hazards of Hazardous Materials for Emergency Responders*. Quincy, MA: National Fire Protection Association.

4 U.S. Department of Transportation. 2012. *Emergency Response Guidebook*. Washington, DC: U.S. Department of Transportation.

Chapter 16
Infrastructure Emergencies

Chapter Objectives

- Identify the typical categories of infrastructure emergencies.
- Identify the incident management priorities for infrastructure emergencies.
- Discuss the operational tactics utilized at infrastructure emergencies.
- Discuss the safety considerations associated with infrastructure emergencies.
- Identify the emergency service agencies that respond to infrastructure emergencies.
- Identify the resource requirements associated with infrastructure emergencies.
- Discuss the story elements associated with media coverage of infrastructure emergencies.

An integral aspect of contemporary life as we know it is the infrastructure system that underpins most aspects of modern society. While the nature and complexity of this infrastructure varies from community to community, with more sophisticated infrastructures typically existing in urban areas in comparison to rural locations, individuals rely on a fully functioning infrastructure to support daily life, work, and travel; organizations likewise count on this infrastructure to support organizational processes and activities.

Incidents that occur on highways and other roads that comprise our nation's transportation infrastructure are considered in a number of related chapters, with accidents and vehicle fires covered in chapters 9 and 14, respectively. Chapters 12 and 15 consider hazardous materials fires and other hazardous materials incidents that may occur on a roadway or highway. Weather-related emergencies and natural disasters, which have the potential of disrupting numerous infrastructure components in addition to roadways, are discussed in chapter 20.

The utilities that provide essential services to a community represent a major component of a community's infrastructure and serve as a primary focus of this chapter. Additional elements of a community's transportation infrastructure are also considered in this chapter, as are selected infrastructure components within buildings, such as elevators. The focus of this chapter is on infrastructure emergencies, or occurrences that disrupt the normal functioning of essential elements of the infrastructure that are taken for granted in contemporary society but are capable of bringing a community to a standstill when their availability is disrupted.

Typical Incidents

While there will be times that a fire department is dispatched to an infrastructure emergency such as downed power lines, a transformer fire, a gas main leak, or a water main break, it is not unusual that the fire department may be dispatched to another type of call and discover the infrastructure emergency while en route or upon arrival on the incident scene. This reality is similar to a fire department discovering that an emergency incident that it responded to, such as a motor vehicle accident, involves a vehicle that was transporting hazardous materials. Likewise, based on a thorough size-up, it could be determined that the damage caused by the vehicles involved in an accident resulted in downed power lines, a ruptured or leaking gas line, or a broken water main. These possibilities once again reinforce the importance of conducting a comprehensive size-up upon arrival on an incident scene.

Various other incidents, including building and equipment fires, could likewise compromise the integrity of the infrastructure in terms of utilities—electric, gas, sewer, telephone, or water—or rail transportation in close proximity to the occurrence of the original emergency. Rail service, whether by train, subway, elevated train, or light rail, can be interrupted due to train tracks being blocked as a result of a power outage or due to signal problems on the rail line. Such delays, cancellations, and shutdowns can affect commuter, regional, and national passenger service, as well as freight trains given that the same tracks are frequently used for various transportation conveyances.

A majority of infrastructure emergencies relate to the utilities that service and provide the lifeblood of a community: electric, gas, sewer, telephone, and water (fig. 16–1). Electrical emergencies include downed electrical power lines, electrical substation fires, electrical vault fires, transformer fires, and other electrical utility emergencies. While most electric utilities utilize grid systems that incorporate equipment redundancy, many of these emergencies result in power outages. Organizations with mission-critical power needs, such as healthcare facilities, radio and television broadcast stations, and many businesses, have prepared for such power interruptions through the installation of backup emergency generators with automatic transfer switching equipment that activates the generator to power the facility automatically upon the loss of electrical power from the utility company (fig. 16–2). A consequence of a power outage is often the activation of numerous automatic fire alarms as well as the stranding of people trapped in elevators at the time of the loss of power. The rescue of passengers trapped in an elevator is discussed in chapter 17.

- Downed electrical power lines
- Electrical substation fire
- Electrical utility emergency
- Electrical vault fire
- Elevator emergency
- Gas line/main break
- Manhole emergency
- Power outage
- Subway emergency
- Transformer fire
- Water main break

Fig. 16–1. Typical incidents: infrastructure emergencies

Fig. 16–2. Emergency generator. (Courtesy of M. Grant Everhart.)

Gas mains and distribution equipment represent another category of potential infrastructure emergency. Leaks, breaks, or ruptures that result in the release of gas into the environment, whether a structure or the atmosphere, can be extremely dangerous based on the potential for ignition or explosion of these hazardous materials, as discussed in chapters 12 and 15. When utilities run underground, the potential for manhole emergencies resulting from fire or explosion exists. Water main breaks can also disrupt a community's infrastructure, resulting in flooded roadways, basements, and sinkholes, and presenting additional challenges in terms of the water coming in contact with charged electrical circuits and causing pilot lights to be extinguished in gas-fired equipment such as water heaters. Infrastructure emergencies can also disrupt traditional land-based telephone service as a result of damage to communication lines or equipment. This is fortunately less of a problem today given the extensive availability and use of more advanced telecommunications devices.

While these various infrastructure emergencies can occur as isolated and independent events, as in the case of a transformer fire on a utility pole, the same transformer fire could instead be just one aspect of the damage or aftermath of an accident, natural disaster, or weather event. Similarly, all of the infrastructure emergencies considered in this chapter could be independent events, or the consequence of one of the other emergencies discussed throughout this book.

Incident Priorities

As with all incidents, the priorities in the management of an infrastructure emergency, in priority order, are (1) life safety, (2) incident stabilization, and (3) property conservation.

Operational Overview

Infrastructure emergencies can represent a fairly significant number of fire department responses in some communities. This is particularly true in communities serviced by an aging infrastructure that is increasingly subject to electrical equipment failures and/or leaks or breaks in gas, water, or sewer lines. The agencies dispatched to such incidents, while determined by the specific dispatch protocols of the jurisdiction, will more times than not involve the fire department in the initial response, along with appropriate utility companies based on the nature of the reported incident. Upon arrival on the scene, fire department officers will perform a thorough size-up of the situation in the interest of determining what has happened, what is happening now, and what is likely to happen. A representative size-up might reveal that a 24-inch water main developed a large break, that a significant quantity of water has flooded the roadway, and that in a fairly short period of time the water runoff will begin to enter the basements of nearby structures, posing additional hazards when it comes in contact with electrical and gas services. As hazardous as a utility emergency involving electricity, gas, or water may be, the associated danger is much greater when more than one of these are involved, examples being electric and water, or electric and gas.

As always, the highest priority in managing these incidents must be ensuring the life safety of all emergency responders and the public. Appropriate control zones must be established and enforced throughout the incident. Both traffic and crowd control can be extremely important at these incidents, as is evacuating individuals from areas that present a danger for harm. Law enforcement agencies can play an important role in these protective actions. All actions taken by the fire department and other emergency response personnel should be based on a comprehensive size-up and evaluation of the incident. The priority of life safety must drive all strategies and actions on the incident scene. All operations should be conducted in accordance with the department's established policies and procedures. Personnel accountability should be ensured throughout the incident, with all fire department personnel, apparatus, and equipment staging an appropriate distance from the hazards, which in many cases may involve a significant distance.

In the majority of infrastructure emergencies the role of the fire department may involve a nonintervention strategy, where it ensures the life safety of all on or near the incident scene while waiting for utility company personnel with the necessary training and equipment to arrive on the scene and resolve the emergency (fig. 16–3). While to the uninformed observer it may appear that the fire department is just standing around doing nothing, were fire department personnel to take things into their own hands and attempt to perform tasks beyond which they are properly qualified and equipped to perform, the consequences could be tragic. While there will be times that utility representatives may request that the fire department remain on the scene for a period of time until the incident is stabilized, in many cases the incident will be turned over to the utility company so fire and emergency service units can return to service and availability to handle other calls. This is particularly true in situations where a community suffers significant storm damage from a weather event resulting in many infrastructure emergencies that cause emergency calls to be "stacked," awaiting dispatch based on the priority associated with the nature of each call.

As with most incidents, exposures can represent a significant issue in the management of infrastructure emergencies. It should be noted that exposures are a two-way street with respect to infrastructure emergencies in that while an infrastructure emergency will often involve potential hazards or damage to adjacent vehicles, buildings, and other property, the reverse can be true as in the case where another type of emergency incident can yield exposures that involve infrastructure components such as electrical, gas, or water lines. Thus, a downed electrical power line could fall on a vehicle or building, a severed gas line could fill a building with gas, a vehicle involved in an accident could damage utility lines, or a building fire could compromise the integrity of the various utility lines within the building. While the fire department may find itself in a situation where certain exposures are in need of protection, these actions should only be undertaken based on a determination that there are no life safety risks associated with their performance.

Fig. 16–3. Utility company personnel handling infrastructure emergency. (Courtesy of David Paul Brown, Montgomery County Department of Public Safety.)

Safety Considerations

Infrastructure emergencies, by their nature, have the potential of presenting major life safety hazards to both the public and emergency responders. Extreme caution must therefore be constantly exercised when emergency responders are operating on the scene of an infrastructure emergency. It must be recognized that in addition to calls actually dispatched as infrastructure emergencies, many other routine incidents may escalate to the involvement of dangerous utilities that are capable of rendering significant harm to emergency responders and the public. It is therefore imperative that, based on a comprehensive size-up of the incident, appropriate strategic goals and tactical objectives be determined. Necessary and appropriate protective control zones should be established in a timely manner and enforced until such time as the associated hazards have been addressed and eliminated. Evacuations of those within hazard areas may be necessary to ensure their safety. Traffic and crowd management are of utmost importance at many infrastructure emergencies. Personnel accountability must be maintained throughout the duration of the incident.

Given the potential for harm that often exists at these incidents, the appointment of an ISO is essential. The individual assigned this important role is charged with the responsibility of monitoring safety issues and advising the incident commander of safety concerns that should be considered in the development and implementation of strategic goals and tactical objectives. Inherent hazards associated with infrastructure emergencies may include the limited access and ventilation associated with confined spaces, potentially hazardous atmospheres, electrical hazards from energized electrical lines and equipment, mechanical hazards, explosive atmospheres, hazardous materials, and ignition sources.

In the majority of situations, a determination will be made to follow a nonintervention strategy of protecting emergency responders and the public while waiting for utility company personnel to arrive and take responsibility for resolving the emergency situation. While once it has been confirmed that involved utilities have been shut off, emergency response personnel may be deployed to handle normal tasks such as fire extinguishment or performing a rescue, these routine tasks should not be initiated until the utility company has confirmed that it is safe.

The above guidance is relevant and prudent for all types of incidents that involve utilities and other infrastructure. Therefore, emergency responders should never operate on rail tracks, whether extinguishing a train fire or grass fire on the tracks or performing a rescue following a train crash or derailment, without first ensuring that the power to the tracks has been de-energized and train traffic has been officially suspended.

The essence of ensuring life safety while operating at an infrastructure emergency involves understanding, respecting, and enacting the appropriate roles of all involved agencies. The fire department is not responsible for performing tasks that fall within the expertise and responsibility of the community's utility companies. In many situations the role of emergency responders should be limited to ensuring life safety through establishing appropriate control zones and perimeters and conducting necessary evacuations. The effective, efficient, and safe resolution of an infrastructure emergency requires a cooperative effort on the part of emergency responders and their utility company counterparts.

Resource Requirements

The typical response to an infrastructure emergency will include the fire department and law enforcement. While the necessary apparatus and equipment will be based on the nature of the incident, the apparatus typically dispatched to infrastructure emergencies will include engine companies and rescue companies. The typical tools used by firefighters at infrastructure emergencies include chemical foam generation and application equipment, fire extinguishers, gas detectors, generators, portable equipment, portable pumps, positive pressure ventilation fans, and traffic control equipment.

Media Coverage of Infrastructure Emergencies

The public interest in news coverage of an infrastructure emergency will vary based on the nature and scope of the incident, as well as its impact on the community. As with other types of emergency incidents, the likelihood of inclusion in news coverage will be based on the extent of damage involved and whether there were any resulting casualties. The instance of a downed electrical power line serves to illustrate the newsworthiness of such incidents. Much like the well-known analogy that if a tree were to fall in the forest no one would hear it, were an energized electrical line to fall in that same forest or other remote area, no one would likely hear about it since it would not receive news coverage. Were that same downed wire to ignite the surrounding forest or wildlands and consume many acres or injure or kill someone, it would more than likely be judged as being newsworthy and would receive news coverage. Likewise, if a fallen electrical wire caused a vehicle or building fire, it too would likely be deemed newsworthy.

As with all news reporting, stakeholders expect that coverage of infrastructure emergencies will be accurate, comprehensive, professional, and timely. In addition to satisfying their interest or curiosity of community events, these news stories often play a role in enhancing the public's understanding of the hazards associated with such emergencies and actions that they should take or not take in such situations. Informative coverage of infrastructure emergencies can contribute to the understanding and insights of the public regarding various infrastructure emergencies. The importance of not creating an ignition source in the presence of a gas leak by turning a light switch off or starting a car ignition would be representative learning objectives of related public education sidebar stories. Likewise, the media can provide an understanding of the importance of not touching or attempting to move any live wire and that even telephone and cable television lines can conduct much higher currents than normal should they come in contact with higher voltage electrical lines.

In addition to reporting on the specifics of an infrastructure emergency, the news media often assume an important public information role within a community from the standpoint of informing their readers, listeners, or viewers of the impact of an infrastructure emergency. This is particularly important in situations where major weather events or natural disasters have compromised significant portions of the utilities servicing a community. Information on the extent of power outages and estimates provided by utility companies as to when the power is likely to be restored is sought-after information that the news media can relay in a timely manner to their audience. In the event of a gas emergency, the news media can provide information regarding the nature and extent of the emergency, the areas affected, and precautions that the public should take with respect to gas emergencies. Residents suffering the impact of flooding resulting from a broken water main will likewise look to their news station to learn about the occurrence and the current status and actions being taken to stop the leak and the corresponding flooding that their residences may be experiencing. Comprehensive news coverage, extended over hours or days, will often be provided in the case of major weather events or natural disasters that result in the extensive disruption or destruction of a community's or region's infrastructure. An important aspect of news coverage of these types of incidents is the provision of information on road closures and detours.

Infrastructure emergencies serve to illustrate the importance of the cooperation and collaboration advocated throughout this book. This is the case both in the management of these incidents and in the associated news coverage. The cooperation of the public in the timely reporting of these emergencies, having patience if it takes time to resolve an emergency situation and exercising the necessary restraint not to increase the hazards of a situation through their ill-advised actions or behaviors, are essential components of the effective, efficient, and safe resolution of infrastructure emergencies. The news media can play an integral public information role toward this end. A second component of this cooperation involves the working relationship between emergency responders and their counterparts who respond to these incidents on behalf of the utility companies, wherein each respects the other's responsibilities, capabilities, and qualifications and acts in accordance with that understanding. Last, but certainly not least, is the cooperative working relationship between the fire department and media representatives that ensures that the coverage of these often extremely dangerous situations fully meets the expectations of all interested stakeholders in terms of providing necessary information, education, and instructions regarding the incident(s).

The job aids in figures 16–4 and 16–5 are designed to serve as resources for fire department and media personnel. Figure 16–4 provides an overview of incident management of infrastructure emergencies, while figure 16–5 provides guidance with respect to media coverage of these incidents.

Incident Management: Infrastructure Emergencies	
Typical Incidents	• Downed electrical power lines • Electrical substation fire • Electrical utility emergency • Electrical vault fire • Elevator emergency • Gas line/main break • Manhole emergency • Power outage • Subway/railroad emergency • Transformer fire • Water main break
Incident Priorities	1. Life safety 2. Incident stabilization 3. Property conservation
Operational Tactics*	• Control zones, crowd control, decontamination, evacuation, exposure protection, nonintervention strategy, operate in accordance with standard operating procedures, perimeter control, personnel accountability, review preplans, scene assessment, scene management, scene security, shelter in place, staging, traffic control, utility control, and vapor suppression
Safety Considerations*	• Carbon monoxide, confined space (limited access, limited ventilation, potentially harmful atmosphere), control zones, crowd control, contamination/decontamination, electrical hazards (energized equipment), electrocution, evacuation, explosion, explosive atmosphere, exposure, fatalities, hazardous materials, ignition sources, injuries, mechanical hazards, Mayday (personnel becoming trapped, disoriented, lost, or running out of breathing air), perimeter control, personnel accountability, rehabilitation, respiratory protection, technical rescue, toxic atmosphere, traffic control, utilities, victims, and water-reactive materials
Responding Agencies*	• Fire department and utility companies (electric, gas, sewer, telephone, and/or water)
Resource Requirements	Personnel* • Fire department, emergency medical services, hazardous materials team, law enforcement, and utility companies Personal protective equipment* • Coat, pants, boots, helmet, gloves, eye protection, personal alert safety system (PASS), self-contained breathing apparatus (SCBA), and specialized personal protective equipment (PPE)

Fig. 16–4. Incident management job aid. **Note:* The nature of a particular incident will determine the appropriate operational tactics, safety considerations, responding agencies, and resource requirements.

	Apparatus* • Firefighting apparatus, foam units, rescue apparatus, support apparatus, and emergency medical units Equipment* • Air cascade system, fire extinguishers, foam generation and application equipment, gas detectors, generators, portable equipment, portable pumps, positive pressure ventilation fans, and traffic control equipment

Fig. 16–4. Incident management job aid (continued)

	Media Coverage: Infrastructure Emergencies
WHO?	• Property owner? • Number of victims? • Number of individuals displaced? • Number of and extent of civilian injuries? • Number of and cause of civilian fatalities? • Number of and extent of firefighter injuries? • Number of and cause of firefighter fatalities? • Personal information on injured individuals? • Personal information on fatalities? • *Have appropriate notifications been made?* • *Has information on victims been received from a credible and authorized source?* • *Has information been verified for accuracy?* • How and to which medical facilities were victims transported? • Medical condition of victims (initial and present)? • *See chapter 21 for incidents involving civilian injuries or fatalities.* • *See chapter 22 for incidents involving firefighter injuries or fatalities.*
WHAT?	• Nature of incident? • Involved utilities? • Electrical emergency? • Gas emergency? • Sewer emergency? • Telephone emergency? • Water emergency? • Extent of infrastructure interruption? • Estimated length of utility interruption? • Associated hazards?

Fig. 16–5. Media coverage job aid

	• Communities affected? • Businesses affected? • Damage to infrastructure? • Damage to buildings or property? • Damage to vehicles? • Responding agencies? • Resource requirements (number and types of personnel, apparatus, and equipment)? • Incident priorities? • Operational tactics? • Safety considerations? • Incident management challenges? • Hazardous materials involvement? • Resulting health or environmental concerns? • Evacuation or shelter in place? • Location of evacuation shelters? • Number of businesses displaced? • Road closures? • Traffic detours?
WHEN?	• Day of week? • Time of day? • Day/night?
WHERE?	• Incident location? • Municipality or neighborhood? • Nearby landmarks (roads, buildings, etc.)? • Impacted area? • Electric utility? • Gas utility? • Sewer utility? • Telephone utility? • Water utility?
WHY?	• Factors that contributed to the incident? • Contributing human factors? • Weather-related factors?
HOW?	• What caused the incident? • Area where the emergency originated? • Did the incident cascade or spread?

Fig. 16–5. Media coverage job aid (continued)

Sidebar Story Ideas	• Community infrastructure • Downed electrical wires • Electrical emergencies • Electrical power restoration following power outages • Electrical utility emergencies • Elevator emergencies • Evacuation during emergencies • Gas emergencies • Highway detour systems • Power outages • Safety precautions during infrastructure emergencies • Sheltering in place • Water emergencies • Weather-related emergencies • When to call utility companies • Winter emergencies
Resources	• National Fire Protection Association (NFPA) www.nfpa.org • National Institute for Occupational Safety and Health (NIOSH) www.niosh.gov • Occupational Safety and Health Administration (OSHA) www.osha.gov • United States Fire Administration (USFA) www.usfa.fema.gov

Fig. 16–5. Media coverage job aid (continued)

Chapter Questions

1. Identify the typical categories of infrastructure emergencies.
2. List the incident management priorities in managing infrastructure emergencies.
3. Discuss several operational tactics related to infrastructure emergencies.
4. Identify the safety concerns related to infrastructure emergencies.
5. Identify the emergency service and support organizations and agencies that are dispatched to infrastructure emergencies.
6. Discuss the resource requirements, in terms of personnel, apparatus, and equipment, associated with infrastructure emergencies.
7. Relate and explain the essential story elements of news coverage of infrastructure emergencies.

Chapter 17
Rescues

Chapter Objectives

- Identify the typical categories of rescues.
- Identify the incident management priorities for rescues.
- Discuss the operational tactics utilized at rescues.
- Discuss the safety considerations associated with rescues.
- Identify the emergency service agencies that respond to rescues.
- Identify the resource requirements associated with rescues.
- Discuss the story elements associated with media coverage of rescues.

Rescue incidents often represent a sizable number of the incidents to which a fire department will respond. This is particularly the case where the jurisdiction served by the fire department is such that its attributes have the potential of contributing to the need for the enactment of the various types of rescue discussed in this chapter. This might involve the various types of water rescue that result from residents and visitors participating in recreational activities, as in the case of becoming a victim while swimming in a body of water such as a swimming pool, the ocean, a river, creek, pond, or lake. The fun of a vacation or planned recreational outing can at times quickly take a tragic turn, necessitating the response of fire and emergency service organizations to render timely assistance to victims. Likewise, an unexpected and unfortunate turn of events could place adventurous individuals engaged in a range of outdoor activities including water skiing, scuba diving, snow skiing, hiking, and mountain climbing in harm's way.

While certain incidents will transpire while their victims engage in recreation as related above, fire departments also respond to incidents where individuals have been injured while engaged in work, as in the case of becoming entangled in machinery, whether while working in an industrial setting on a factory assembly line or an agricultural setting while operating farm machinery. Victims likewise fall prey to accidents that necessitate vehicle rescue as they travel for a variety of purposes. Communities that have major roadways, including high-speed divided highways, frequently experience vehicle accidents and the need for associated rescue services.

Typical Incidents

The majority of rescue incidents to which most fire departments are dispatched involve vehicle rescue. As discussed in several earlier chapters, vehicles and their associated challenges, whether in a fire or rescue situation, will vary based on the vehicle(s) involved, the nature of the incident, the location of the incident, whether the incident involves fire or hazardous materials as discussed in chapters 12 and 15, respectively, or downed power lines or damaged utilities such as a gas line as discussed in chapter 16. Vehicle rescue incidents range from those where victims are actually trapped by the vehicle and require vehicle extrication and those of lesser severity where the fire department only needs to "pop" a door open to create a pathway for a victim to self-extricate. Most vehicle rescues fall within the resources and capabilities of the local fire department. Review of the diverse range of rescue incidents delineated in figure 17–1 provides an understanding of the challenges that fire departments may face in ensuring their preparedness to address these incidents within the communities that they serve.

Many of the incidents that we examine in this chapter are considered *technical* or *specialized rescue* and typically require specialized teams to assist the fire department in their resolution. Rescue incidents in the workplace include *agricultural rescue* where a farm worker becomes entangled in the operating machinery of a tractor, combine, or other farm machinery or similar machinery extrication when a factory worker becomes a victim of entanglement in a moving conveyor or assembly line. The fact that the involved machinery may continue to operate unless it has certain safety devices, such as a dead man's switch, makes it imperative that involved operating machinery be deactivated by shutting it down in a timely manner so as to prevent further injury to the victim. A related rescue category is that of *elevator rescue*. While these incidents do not involve the traumatic injuries associated with agricultural or *machinery rescue*, to the victims trapped inside an elevator, especially for an extended period of time or if they have existing medical or psychological issues, these incidents can be very stressful.

- Agricultural rescue
- Animal rescue
- Building collapse
- Cave-in
- Confined space rescue
- Drowning
- Elevator rescue
- Firefighter rescue
- High-angle rescue
- Ice rescue
- Machinery rescue
- Rope rescue
- Structural collapse rescue
- Surf rescue
- Swiftwater rescue
- Trench rescue
- Vehicle rescue
- Water rescue

Fig. 17–1. Typical incidents: rescues

Fire departments and other emergency responders who serve communities with recreational attractions are often called upon to conduct associated rescues, as in the case of water rescues when someone falls victim in a swimming pool or natural body of water. A *surf rescue* would be necessitated in the case of an individual drowning while swimming or surfing in the ocean. Whether an individual is engaged in planned recreational activities or simply finds him- or herself in the wrong place at the wrong time, the moving waters resulting from severe weather, a natural disaster, or an infrastructure emergency such as a water main break, can quickly place that person in harm's way triggering the necessity of a *swiftwater rescue*. Whereas many of the recreational accidents discussed above typically occur during the warmer seasons, the need for *ice rescues* may arise in some communities during the winter months when victims become trapped on an ice-covered body of water or fall through the ice.

Some rescue incidents, referred to as *high-angle rescues*, involve the rescuing of victims trapped at elevations—an injured mountain climber, a child who has slid down a cliff or embankment while playing in a rural area, an injured construction worker on a high-rise construction site, a stranded window washer, or a worker trapped and/or injured in a high tree or on a radio transmission tower. The effective, efficient, and safe resolution of many of these incidents will involve the associated techniques and technologies of *rope rescue* (fig. 17–2). Some rescue operations involve *confined space rescue* and *structural collapse rescue*, as in cases of building collapse. Victims can be trapped as a result of natural occurrences such as cave-ins and avalanches. The rescue of victims trapped in an artificial excavation, such as a foundation excavated for a new building or a trench dug during infrastructure or utility installations or repairs, is considered a *trench rescue*.

Fig. 17–2. Firefighters training in rope rigging. (Courtesy of Pennsylvania State Fire Academy.)

As always, the priority in all types of rescue is life safety of the victims of the incident and emergency responders. While the victims of an incident will most often be civilians, such as those trapped during a building collapse, at times the victims may also include firefighters or other emergency responders who become lost, disoriented, injured, trapped, or run out of breathing air while performing their assigned tasks on the scene of an emergency incident. These unfortunate turns of events often result in the declaration of a Mayday by involved personnel and initiate a *firefighter rescue*, wherein the rescuer that has fallen victim now becomes the target of rescue efforts. While most rescue operations are targeted at saving people, there will be times that rescue operations involve rescuing animals, as in the case of the *animal rescue* of a dog trapped on an ice-covered lake or a household pet during a house fire or structural collapse.

Incident Priorities

As with all incidents, the priorities in the management of rescues, in priority order, are (1) life safety, (2) incident stabilization, and (3) property conservation.

Operational Overview

Successful management of any rescue incident begins with performing a comprehensive situational assessment of the incident in the interest of developing and implementing an incident action plan that will contribute to its effective, efficient, and safe resolution. As always, the priority of life safety of emergency responders and the public, in this case the victim(s) who are in need of being rescued, must guide all decisions made and actions taken on the incident scene. Given that victims may have suffered injuries requiring timely medical attention, efficient operations are important but should never come at the expense of compromising the life safety of rescuers or victims. A primary assessment of victims should be conducted as soon as possible, with a further secondary assessment of their injuries and medical condition being conducted when practical, typically subsequent to their removal to a safe area.

The nature and specifics of a particular rescue incident will determine the appropriate operational tactics to resolve the emergency situation. While the scope of this book prevents exhaustive coverage of the operational requirements of the various types of rescue incidents, a number of requisite elements of successful incident management, including the provisions for the safety of both victims and rescuers, are discussed.

Given the technical requirements of many of these incidents, specialized resources, whose personnel have the appropriate training, certifications and equipment to effectively, efficiently, and safely address the operational needs of the incident will usually be requested and utilized. The diversity of rescue operations dictate training and professional certifications in areas such as confined space, high-angle, machinery, rope, structural collapse, and water rescue. *NFPA 1006: Standard for Technical Rescuer Professional Qualifications* serves as the definitive document that outlines the job performance requirements in terms of knowledge and skills for the various technical rescue specialties.[1]

The management of each of these incidents, whether a nontechnical rescue on flat or level ground routinely handled by the local fire department or a technical rescue involving a victim trapped on steep terrain or at a high elevation requiring the services of a high-angle rescue team, demands a thorough consideration of all of the inherent risks and hazards associated with the operational tactics under consideration. An integral aspect of the resulting incident action plan must be ensuring that rescue personnel have an appropriate and safe pathway from which to gain access to and retrieve the victim(s). A variety of tactics, techniques, and equipment will often be required to resolve these emergency incidents, such as the use of personal flotation devices and safety lines during water rescue evolutions, life safety harnesses and safety lines properly manned and secured to anchor points during high-angle rescues, and appropriate harnesses and air supply systems during the conduct of confined space rescue operations.

A challenge that may complicate some of these incidents is the medical condition of the victim, either based on a preexisting medical condition or as a result of injuries suffered during the triggering event that resulted in the rescue. It is imperative that rescue operations be conducted in a manner that protects the incident victim(s) and does not further contribute to their injuries. It may thus be necessary for rescuers to transport not only rescue equipment, such as a Stokes basket or harnesses that will be used by rescuers to "package" the victim before safe removal and transport, but also additional personal protective equipment, such as helmets, that when worn by the victim will offer enhanced protection (fig. 17–3). While the desire will always be to perform rescue operations in a manner that results in a successful outcome, there will be times that it will be determined that the potential to save the victim(s) is not a viable option, thus dictating a change in strategic goals and operational mode from a rescue to a recovery operation, as in the case of a victim of a drowning that is not located for several days or a victim of a structural or trench collapse that is located and found to have succumbed to his or her injuries.

Fig. 17–3. Rescuers "packaging" a victim. (Courtesy of Bob Sullivan.)

Safety Considerations

While the management of all emergency incidents requires the incident commander to develop an informed incident action plan that ensures the safety of emergency responders and has as its foundation sound decision making based on a comprehensive initial and ongoing situational assessment and cost-benefit analysis, technical and specialized rescue incidents especially demand such deliberate development and implementation of strategic goals and operational tactics. The nature of many of these incidents can involve extraordinary hazards when compared with those associated with extricating a victim from a routine vehicle accident. The utilization of an incident command system, including the appointment of a rescue officer and an incident safety officer (ISO), both of whom have the requisite technical knowledge and skills to successfully enact their responsibilities, is a must, as is tracking personnel accountability throughout the incident.

As with all incidents, the involved rescue personnel must have the appropriate training and qualifications for the assignments that they undertake, this need typically being addressed through the utilization of specialized technical rescue teams. It is important to note that while many technical rescue teams are qualified to perform more than one type of technical or specialized rescue, others are not. Examples would be technical rescue teams that specialize in a particular discipline, such as trench rescue, water rescue, or structural collapse rescue. The utilization of such specialized teams enhances the safety of responders and victims in two important ways: through assisting the incident commander in developing a realistic and safe incident action plan and through providing the personnel who are qualified and equipped to perform the necessary supporting operational tactics.

There are some common considerations in ensuring life safety at all rescue incidents, such as preventing the movement of elements of the environment surrounding the victim, whether an automobile during a vehicle rescue, the surrounding earth during a trench rescue, building structural components and furnishings during a structural collapse rescue, or elements of the terrain during a high-angle rescue. Likewise, it will always be imperative that rescue personnel wear the appropriate personal protective equipment for the operational tasks at hand. While the structural turnout gear that firefighters routinely wear serves them well during various types of firefighting, the utilization of this same protective gear during many technical rescue operations would actually compromise their safety, as in adding mass and weight that could actually cause them to drown during certain water rescue incidents. Required personal protective equipment at such incidents would obviously include personal flotation devices, life safety harnesses, and life safety lines.

The provision of a reliable air supply will be essential at many rescue incidents. Personnel from dive rescue teams will utilize appropriate air supply systems. Personnel operating in confined spaces and areas where the atmospheric breathing air is compromised will likewise need a reliable air supply source. Given the length of time it may take to perform certain operational tasks, supplied-air systems that provide a continued source of breathing air may be utilized rather than the limited duration air tanks incorporated in the self-contained breathing apparatus (SCBA) that firefighters usually don during firefighting operations. As always, proper supervision and crew management are imperative in ensuring life safety at these incidents. Both the duration of some of these incidents and the accompanying physical demands on rescue personnel result in the need to provide for the rehabilitation of personnel at many of these incidents. The importance of rehabilitation is illustrated by the fact that the appropriate personal protective equipment ensemble for many rescue disciplines includes the incorporation of a personal hydration system.

Resource Requirements

The resources required to handle a rescue incident vary greatly in accordance with the nature and size of that incident. While most local fire departments will be fully capable of handling a minor automobile accident where the driver or occupants are merely confined and thus only need a door "popped" open, at larger vehicle accidents where there is major entrapment the assistance of mutual aid companies with medium- or heavy-duty rescue trucks may be required. Likewise, the number of vehicles within which occupants are trapped may trigger the dispatch of additional rescue apparatus. The resource needs of the various types of technical rescues will correspondingly result in the dispatch of various specialized apparatus such as boats or a rescue helicopter for water rescues, or various types of heavy construction equipment, such as a crane, in the event of a building collapse. Support apparatus, such as a communications unit, will be dispatched to many of these incidents. In addition to fire and rescue apparatus, the dispatch of aerial apparatus may occur based on the equipment that some of these trucks carry, as well as their ability to serve as a high anchor point in certain rescue operations.

While there are some common tools and equipment such as cribbing, cutting tools, ground ladders, hand tools, pneumatic tools, portable equipment, power extrication tools, power tools, prying tools, and striking tools used in various rescue operations, some of the equipment used at technical rescue incidents may be specific to the needs of one or more types of technical rescue. Representative examples of such equipment would be air monitoring equipment, air supply carts, air supply systems, air bags, anchor systems, gas meters, hydraulic jacks, hydraulic shoring, inflatable boats, line guns, marking equipment, pneumatic shoring equipment, pulley systems, rescue rope, rescue tripods, rigging systems, safety rope, shoring equipment, thermal imaging cameras, and winches. Special *intrinsically safe equipment*, including intercoms, lights, and ventilation fans, will be required in potentially explosive environments where technical rescue operations are conducted. Appropriate emergency medical service units, either basic life support or advanced life support, will be dispatched to rescue incidents. Patient care equipment typically includes patient immobilization devices, rescue litters or Stokes baskets, and transport stretchers or litters.

Media Coverage of Rescues

While many rescue incidents, such as a single-vehicle automobile accident involving limited injuries to the vehicle's occupants or property damage, will be fairly routine for the responding fire department and of limited interest to the news media and their audience, larger and more unique rescue incidents routinely garner interest of the news media and the audiences on whose behalf they cover these breaking stories. The various types of technical rescue discussed throughout this chapter all present stories that are both interesting to cover and have the potential for related informative sidebar stories. In addition to covering a dramatic high-angle rescue of a worker who experienced a medical emergency or traumatic injury while dangling 100 feet in the air, a local news station could incorporate sidebar coverage on the challenges and techniques of high-angle rescue that features the regional technical rescue team. Given the dramatic nature of these stories, the incorporation of visual images into reporting is certainly desirable, with news organizations often capturing these images from both the ground and the air. These stories frequently are picked up by network affiliates and incorporated in their news reporting.

Given the challenges of managing some technical rescue incidents, it will often be the case that during the initial minutes of the incident, the incident commander and other fire department personnel will have limited time to devote to the news media, focusing their attention on assessing the incident and developing and implementing the tactics necessary to effectively, efficiently, and safely resolve the emergency situation. This is another example of the value of utilizing a PIO. Given the technical nature of these incidents, it will be in the best interest of the fire department to, at an appropriate point during the incident, afford news reporters the opportunity to talk to the incident commander and the technical specialists who oversee the tactical rescue operations.

As always, it is imperative that the media not interfere in any manner with incident operations, and while it is desirable to provide a media area and access from which photojournalists can capture the oftentimes award-winning photo that is picked up by the syndicated news agencies or the video segment of the incident that the various television networks run, this must never be the result of allowing the news media to operate from vantage points where their life safety or that of others could be compromised. Once again, these are stories that, through cooperation and collaboration between fire and emergency service organizations and the news media, can often chronicle the great work of those organizations and individuals that provide exceptional fire and emergency services to the communities that they serve, regardless of the nature or challenges of a particular incident.

The two job aids in figures 17–4 and 17–5 are designed to serve as resources for fire department and media personnel. Figure 17–4 provides an overview of incident management of rescues, while figure 17–5 provides guidance with respect to media coverage of these incidents.

Incident Management: Rescues	
Typical Incidents	• Agricultural rescue • Animal rescue • Building collapse • Cave-in • Confined space rescue • Drowning • Elevator rescue • Firefighter rescue • High-angle rescue • Ice rescue • Machinery rescue • Rope rescue • Structural collapse rescue • Surf rescue • Swiftwater rescue • Trench rescue • Vehicle rescue • Water rescue
Incident Priorities	1. Life safety 2. Incident stabilization 3. Property conservation
Operational Tactics*	• Air monitoring, communication, control zones, crowd control, evacuation, extrication, hazard assessment, helicopter operations, landing zones, patient assessment, patient packaging, patient treatment, patient transport, perimeter control, personnel accountability, post-incident analysis, pre-rescue briefing, recovery, rehabilitation, rescue, rigging, rope rescue, scene assessment, scene management, selection of appropriate operational tactics, shelter in place, size-up, stabilization, staging, technical extrication (lowering and raising), traffic control, utility control, and ventilation

Fig. 17–4. Incident management job aid. **Note:* The nature of a particular incident will determine the appropriate operational tactics, safety considerations, responding agencies, and resource requirements.

Safety Considerations*	• Appropriate personal protective equipment, atmospheric (air) monitoring, backup teams, breathing air, building collapse, building construction, building hazards, carbon monoxide, communication, confined spaces (limited access, limited ventilation, potentially harmful atmospheres), control zones, crowd control, evacuation, explosive atmospheres, hazards, hazardous materials, heat stress, high elevations, immediately dangerous to life and health (IDLH) atmospheres, ignition sources, incident safety officer (ISO), injuries, intrinsically safe equipment, landing zones, life safety (rescue) harnesses, lifelines, Mayday (becoming disoriented, lost, trapped, or running out of breathing air), perimeter control, personnel accountability, rehabilitation, scene assessment, scene management, stabilization, terrain, utilities, vehicular traffic, and victims • The utilization of appropriate personal protective equipment (PPE) based on the nature of the incident and assigned tasks is essential.
Responding Agencies*	• Fire department, emergency medical services (basic life support and advanced life support), law enforcement, and technical and specialized rescue teams
Resource Requirements	Personnel* • Fire department, emergency medical services, technical and specialized rescue teams, and law enforcement • Essential positions at technical and specialized rescues include incident commander, incident safety officer (ISO), and rescue officer. • The assignment of backup teams to support primary rescue teams is common within technical and specialized rescue. Personal protective equipment* • The appropriate personal protective equipment will be based on the nature of the incident and the assigned operational tasks. • Often rescuers will provide separate personal protective equipment that victims will wear during rescue operations to protect them from additional injuries. • Boots, eye protection, gloves, jump suit, life safety (rescue) harness, supplied air device/system, personal flotation devices, hydration system, accessory pouch, and radio harness Apparatus* • Fire apparatus, rescue apparatus (light-, medium-, and heavy-duty), emergency medical units (basic life support and advanced life support), specialized apparatus (boats, helicopters), and support apparatus

Fig. 17–4. Incident management job aid (continued)

	Equipment* • The potential hazards of many environments in which technical rescue operations are conducted require that intrinsically safe equipment be utilized in the interest of preventing ignition sources. • Air monitoring equipment, air supply carts, air supply systems, air bags, anchor systems, cribbing, cutting tools, gas meters, ground ladders, hand tools, hydraulic jacks, hydraulic shoring, inflatable boats, intrinsically safe intercoms, intrinsically safe lights, intrinsically safe ventilation fans, lifeline rope, line guns, marking equipment, patient immobilization devices, pneumatic shoring, pneumatic tools, portable equipment, power extrication tools, power tools, prying tools, pulley systems, rescue litters/baskets, rescue rope, rescue tripods, rigging systems, safety ropes, shoring equipment, stretchers and litters, striking tools, and winches

Fig. 17–4. Incident management job aid (continued)

Media Coverage: Rescues	
WHO?	• Number of victims? • Witnesses to incident? • Party reporting incident? • Number of and extent of civilian injuries? • Number of and cause of civilian fatalities? • Number of and extent of firefighter injuries? • Number of and cause of firefighter fatalities? • Number of trapped persons? • Number of stranded persons? • Number of missing or unaccounted for persons? • Personal information on injured individuals? • Personal information on fatalities? • Personal information on trapped, stranded, or missing persons? • *Have appropriate notifications been made?* • *Has information on victims been received from a credible and authorized source?* • *Has information been verified for accuracy?* • How and to which medical facilities were victims transported? • Medical condition of victims (initial and present)? • *See chapter 21 for incidents involving civilian injuries or fatalities.* • *See chapter 22 for incidents involving firefighter injuries or fatalities.*

Fig. 17–5. Media coverage job aid

WHAT?	• Nature of incident? • Specific type of rescue? • Responding agencies? • Resource requirements (number and types of personnel, apparatus, and equipment)? • Technical and specialized resources? • Assisting agencies? • Incident priorities? • Strategic goals? • Operational tactics? • Safety considerations? • Incident management challenges? • Hazardous materials involvement? • Involvement of utilities? • Utility companies? • Situational hazards? • Weather at time incident occurred? • Current weather? • Forecasted weather? • Temperature? • Humidity? • Wind direction? • Wind speed? • Wind gusts? • Type of precipitation? • Amount of rainfall? • Amount of snowfall? • Flooding? • Flood stage? • Estimated dimensions of trench? • Location of confined space? • Estimated dimensions of confined space? • Depth of standing water? • Depth of moving water? • Speed of water movement? • Building where structural collapse occurred? • Building dimensions? • Number of building floors? • Location of collapse within building? • Areas of building involved in collapse? • Type of structural collapse?

Fig. 17–5. Media coverage job aid (continued)

	• Evacuation or shelter in place? • Road closures? • Traffic detours?
WHEN?	• Day of week? • Time of day? • Day/night? • Time incident occurred? • Expected duration of incident? • Time(s) rescue(s) were completed? • Time rescue efforts were suspended? • Time rescue efforts will resume? • Time rescue efforts were terminated? • Time when operations transitioned from rescue to recovery mode? • Time(s) victim(s) were recovered?
WHERE?	• Location of incident? • Municipality or neighborhood? • Nearby landmarks (roads, buildings, etc.)? • Past related incidents in area?
WHY?	• Factors that contributed to the incident? • Contributing human factors? • What was victim(s) doing when incident occurred? • Environmental factors that contributed to the incident? • Weather-related factors?
HOW?	• Official determination of incident cause(s)? • Is the incident being investigated? • Agencies involved in investigation? • Are charges being brought against responsible parties?
Sidebar Story Ideas	• Agricultural rescue • Animal rescue • Building collapse • Cave-ins • Confined space rescue • Drownings • Elevator rescue • Firefighter rescue • High-angle rescue • Ice rescue • Machinery rescue • Surf rescue • Swiftwater rescue

Fig. 17–5. Media coverage job aid (continued)

	• Technical rescue teams • Trench rescue • Urban search and rescue (USAR) teams • Vehicle rescue • Water rescue
Resources	• Federal Emergency Management Agency (FEMA) www.fema.gov • National Fire Protection Association (NFPA) www.nfpa.org • National Transportation Safety Board (NTSB) www.ntsb.gov • U.S. Department of Transportation (DOT) www.dot.gov • United States Fire Administration (USFA) www.usfa.fema.gov

Fig. 17–5. Media coverage job aid (continued)

Chapter Questions

1. Identify the typical categories of rescues.
2. List the incident management priorities in managing rescues.
3. Discuss several operational tactics related to rescues.
4. Identify the safety concerns related to rescues.
5. Identify the emergency service and support organizations and agencies that are dispatched to rescues.
6. Discuss the resource requirements, in terms of personnel, apparatus, and equipment, associated with rescues.
7. Relate and explain the essential story elements of news coverage of rescues.

Notes

1 National Fire Protection Association. 2008. *NFPA 1006: Standard for Technical Rescuer Professional Qualifications*. Quincy, MA: National Fire Protection Association.

Chapter 18
Searches

Chapter Objectives

- Identify the typical categories of searches.
- Identify the incident management priorities for searches.
- Discuss the operational tactics utilized at searches.
- Discuss the safety considerations associated with searches.
- Identify the emergency service agencies that respond to searches.
- Identify the resource requirements associated with searches.
- Discuss the story elements associated with media coverage of searches.

The frequency with which fire and emergency service organizations face the challenges of conducting search operations will vary from one department to another, as will the specific nature of these searches. These searches can typically be categorized as involving searching for lost persons in buildings, underground, in bodies of water, or on land. While this chapter focuses on land searches for missing individuals, there will be times when these searches may also involve water if the search area includes bodies of water. A number of the earlier chapters considered incidents that involve searching for and rescuing individuals who are lost or otherwise in harm's way. Chapter 17, for example, discussed searching for civilians or firefighters who are lost, trapped, or disoriented within buildings; those who were last seen in or near a body of water where an individual may have drowned; or the victims of a cave-in as in the case of a mine or tunnel collapse. The goals in all of these incidents include locating the involved person(s), rescuing and removing them to a safe location, and ensuring that they receive necessary medical attention. Unfortunately, as previously discussed, the outcome in some of these incidents involves a body recovery, rather than a successful rescue. There are three possible outcomes to a land search: the subject walks out, the subject is located and brought out by rescuers, or the body of the deceased subject is located and recovered.

Typical Incidents

While the nature of a particular search operation will vary based on the circumstances involved in searches for lost individuals being conducted on land, in water, underground, or in buildings, as listed in figure 18–1, a common denominator in all of these searches is that they involve lost persons. A second common attribute of these incidents is that they all represent emergencies, given the established priority of life safety and the potential vulnerability for search subjects to suffer additional harm if not located and rescued in a timely manner. A variable that will likely differ based on the circumstances involved is whether the person has been accounted for and his or her relative location is known, as in the case of a trapped firefighter who has declared a Mayday indicating the conditions of his or her emergency and location within the building. Information on the operating location of each firefighting crew within a building would likewise be available through the utilization of a personnel accountability system. Likewise, mining personnel trapped in a mine collapse would typically be able to summon help and inform rescue personnel of the nature of their situation and their location. Were witnesses in the water or on land to observe a swimmer or boater going underwater and not coming back up, this too would validate that an individual was in trouble and had potentially drowned and the location at which they were last seen. While each of these situations involves incident management challenges, as discussed in the previous chapters, they typically yield fairly definitive information and insights regarding the subject of a search, their situation, and their location.

- Search for lost person(s) (building)
- Search for lost person(s) (land)
- Search for lost person(s) (underground)
- Search for lost person(s) (water)
- Subject(s) of search (children, adults, the elderly)
- Medical condition or psychological state of search subject(s)
- Activity at time subject(s) went missing
- Situational and environmental factors
- Weather conditions

Fig. 18–1. Typical incidents: searches

Unfortunately, this is often not the case in conducting land searches for missing persons, the subject of this chapter. The one thing that is known is that the involved person(s) cannot be located or accounted for, with the details surrounding the disappearance often being extremely sketchy or unknown. These situations range from a person who went for a walk and did not return to a person engaged in outdoor activities such as climbing, fishing, hiking, or hunting not returning by an expected time. In some cases these activities will be planned in advance and communicated to friends or relatives, while in others they may be spontaneous.

There are a number of factors that will determine how far a person may travel that will influence the management of a search, including the subject's age, medical condition, psychological state, and experience. The age of search victims range from young children who might wander aimlessly after becoming lost, to elderly adults, including those suffering from Alzheimer's or medical conditions that would affect their judgment and ability to travel a distance. Elderly individuals may have hearing impairments that may make it difficult for them to hear search personnel. The subjects of a search, regardless of age, could be suffering from a mental impairment or could be despondent, thus leading to their going missing and increasing the challenges associated with locating them in a timely manner.

The activity that the subject of the search was engaged in will likewise play a role in where he or she might be found by search teams. Situational and environmental factors, such as the time of the day, season of the year, terrain of the search area, and the weather will all be factors that may play a dynamic role in the operational requirements and associated challenges of conducting a search. Likewise, the size of the involved search area will determine the necessary resources as well as the time required to conduct the search, with some searches spanning a number of operational periods and occurring over a number of days.

Incident Priorities

As with all incidents, the priorities in the management of searches, in priority order, are (1) life safety, (2) incident stabilization, and (3) property conservation.

Operational Overview

There will be times that a person has purposely decided to travel somewhere and has his or her full faculties to find the way back, as in the case of an older person deciding to go for a long walk on a nice day, something that he or she might not characteristically do. This would typically be a much more disturbing development if the involved individual was a child or an individual of any age suffering from a condition that would put him or her at risk under these circumstances. While there will be times that individuals who have wandered off will return on their own and the search will be terminated, this will often not be the case, and it must be assumed that the longer the person is missing, the more likely he or she may be subject to harm.

Searches must therefore be planned and conducted with the priority of all emergency incidents where the life safety of the public is at risk. A thorough situational assessment is imperative before commencing search operations. These incidents require detailed planning and coordination that begins with the identification of the lost person(s), along with all relevant information that will contribute to a successful search, including name, age, experience, physical and mental condition, and the clothing worn when last seen. Search personnel will be interested in gaining as much relevant information as possible, including a photograph of the missing individual. If the activity that the lost person had planned or was engaged in is known, that information can prove extremely useful in determining an appropriate search area and techniques.

It will ideally be possible to determine the last location where the subject was seen and the direction of travel through interviewing friends, family, acquaintances, or witnesses, as well as reviewing any records that documented the planned activity during which the person went missing. All of the aforementioned information will provide important guidance that those managing the search will use to identify the search area and prioritize search activities within it. Once a search area has been mapped out, this should be communicated to all search personnel during a pre-search briefing, as should related information on the subject of the search. A search perimeter should be established and enforced in the interest of limiting entry into the search area, and more particularly preventing the subject of the search from wandering out of the search area without being observed. As related above, it is not uncommon for lost persons to find their way out of the search area, thus confinement techniques are essential at these incidents. Once the search area has been cordoned off through confinement, search teams may use various tactics to attract the subject of the search to come to them, as well as engage in search techniques designed to thoroughly search the defined search area using manned search teams, canine search dogs, and aircraft surveillance of the search area (fig. 18–2).

The earlier mentioned attributes of the lost person will influence his or her behavior and provide insights that experienced rescuers will utilize to ensure effective, efficient, and safe search operations. The personal information gathered prior to commencing the search will provide important guidance in the planning and execution of the search. Likewise, the circumstances of the person going missing, the terrain and hazards associated with the search area, the time of day and season of the year, and the current and forecasted weather will serve as important determinants in the development of an appropriate search strategy and tactics (fig. 18–3).

Fig. 18–2. Canine search team

Fig. 18–3. Search briefing. (Courtesy of Bob Sullivan.)

During search operations, search personnel will interview persons they encounter in the interest of determining if they have seen the missing person(s) and gathering any pertinent information they can provide. They will also search for physical clues including discarded clothing or other items, as well as footprints. Search processes will focus on finding clues that lead to locating the subject(s) of the search. The importance of determining an appropriate search area based on known information and establishing and enforcing necessary perimeter controls bears repeating. This facilitates the process of searching for and locating the subject while searching within the search area, or locating the victim as he or she crosses the search area perimeter while walking out of the search area.

Provisions must similarly be made to provide for the rescue of the subject(s) of the search subsequent to them being located, as well as to ensure that they receive required medical attention in a timely manner. Once found, these individuals must be assessed, treated, and transported, as necessary.

Safety Considerations

Many of the earlier mentioned circumstances under which a search may take place will present potential safety issues that those emergency personnel involved in the search may face. While to the naïve observer, it might appear that search operations should be fairly easy to plan and conduct, nothing could be further from the truth. As with other technical areas, such as hazardous materials and technical rescue, search operations are a specialty field that requires that these incidents be managed by and staffed with individuals who have the necessary training, qualifications, and experience to enact these important responsibilities in an effective, efficient, and safe manner. This is essential to ensure the life safety of both those involved in conducting the search and those for whom they are searching.

Scene assessment and management are essential throughout the management of search operations, as is ensuring that all search personnel are equipped with the necessary communication capabilities to maintain contact with those responsible for managing the incident.

Personnel accountability is essential at these incidents, particularly given the reality that the terrain and other hazards associated with the search area and the weather could pose the potential for harm to those emergency responders conducting the search. Establishing necessary control zones that control crowds and vehicular traffic is also important in ensuring the life safety of personnel engaged in search operations. As always, a qualified incident safety officer (ISO) should be assigned to monitor and address issues related to the life safety of response personnel and the public.

These incidents can be long-duration incidents that can subject those involved in search operations to extreme temperatures, which may lead to heat stroke or heat exhaustion. It is imperative that search personnel be provided appropriate personal protective equipment consistent with their assigned tasks and the environment in which they are working. Rehabilitation is likewise a must at these incidents. While a successful outcome is always hoped for, in cases where a person is not found or a deceased body is located, provision will usually be made to conduct a critical incident stress debriefing in the interest of assisting emergency personnel in dealing with the emotional or psychological aspects associated with such tragic events.

Resource Requirements

While some more limited search operations may be conducted by a single emergency response agency, it is typical that search operations will involve resources, including personnel, from a number of organizations or agencies. In addition to the fire department, emergency medical services, and law enforcement, these incidents will frequently trigger the response of specialized teams with expertise in planning and conducting effective, efficient, and safe search operations. In addition to furnishing specialized equipment for use in mapping a search area, tracking missing persons, and accounting for search personnel, these technical teams will bring a highly trained cadre of human trackers, as well as canines trained in trailing and tracking of lost individuals. Depending on the location and terrain, trained personnel on horses may also join the search.

Additionally, a variety of support resources, such as communication and rehabilitation units, will usually be deployed to these incidents. Both helicopters and fixed-wing aircraft will often be required in the management of these incidents. A variety of specialized vehicles, including all-terrain vehicles, mountain bikes, and snowmobiles may be utilized based on the specifics of a particular search incident. By their very nature, these incidents can be extremely labor-intensive, particularly if a large initial search area is established or the decision is made to later expand the initial search area. These are therefore incidents that will often result in a sizeable assemblage of resources secured through both automatic and mutual aid, and at times outside aid.

Media Coverage of Searches

As always, the involvement of people in harm's way makes incidents involving the search for missing persons news stories of interest to the news media and their audiences. These incidents involve human interest stories with which the public will often empathize, particularly if the incident involves a somewhat defenseless, vulnerable individual based on age, medical condition, diminished mental status, or other unfortunate limitations that further place him or her in harm's way. The public will be interested in learning, through news coverage of the emergency incident, what happened, who it happened to, and what is being done to promptly locate the individual and get him or her the necessary help.

These are stories where the residents of a local community or wider region will often come together in their shared concern for the missing person and often pray for their safety as they anxiously wait to hear that the person has survived the ordeal and has been found. These are, once again, news stories where it is imperative that reporting be not only timely, but also accurate, credible, and professional. Likewise, they are stories where responsible reporting demands that personal information on the missing person(s) not be released to or through the news media until such time as appropriate notifications have been made. The importance of verifying the accuracy and credibility of such information is apparent. These are incidents where the utilization of a PIO will contribute to information dissemination and media coverage that meets stakeholder expectations. A role of the PIO at these incidents will often be to arrange for news reporters to speak with appropriate representatives from the emergency response agencies involved in the incident, such as the incident commander, the search team leader, or authorized law enforcement personnel.

These may be incidents where those responsible for managing the incident may utilize the news media to disseminate information about the missing person in the interest of receiving information from the public that might prove instrumental in locating that person. This is another example of how collaboration between emergency management and response agencies and their counterparts in the news media can contribute to successful incident management, as well as desirable incident outcomes.

The job aids in figures 18–5 and 18–6 are designed to serve as resources for fire department and media personnel. Figure 18–4 provides an overview of incident management of searches, while figure 18–5 provides guidance with respect to media coverage of these incidents.

Incident Management: Searches	
Typical Incidents	• Search for lost person(s) (building) • Search for lost person(s) (land) • Search for lost person(s) (underground) • Search for lost person(s) (water) • Subject(s) of search (children, adults, the elderly) • Medical condition or psychological state of search subjects • Activity at time subject(s) went missing • Situational and environmental factors • Weather conditions
Incident Priorities	1. Life safety 2. Incident stabilization 3. Property conservation
Operational Tactics*	• These incidents involve locating lost persons, rescuing and removing them to a safe location, and ensuring that they receive necessary medical attention. • Possible outcomes of a land search are subject walks out, subject is located and brought out by rescuers, or the body of a deceased subject is located and recovered. • Identification of lost person(s), gather relevant personal information on subject, secure any documentation of planned activities, secure photograph of subject, conduct relevant interviews (family, friends, acquaintances, witnesses), identify location where subject was last seen, conduct situational assessment, determine search area, conduct pre-search briefing, establish search area perimeter, confine subject to search area, monitor search area perimeter for "walkout," use control zones in crowd and traffic control, search for physical clues, segment and prioritize search area, locate missing subject, rescue and remove subject to safety, and provide necessary medical treatment
Safety Considerations*	• Ensuring the safety of search personnel and the subject(s) of a search requires that those individuals managing search operations, as well as those conducting the search, have the requisite training, qualifications, and experience to effectively, efficiently, and safely enact their responsibilities. • Search management, similar to hazardous materials and technical rescue, represents a specialty area in which personnel must be properly trained prior to assuming related roles at an emergency incident involving a search.

Fig. 18–4. Incident management job aid. **Note:* The nature of a particular incident will determine the appropriate operational tactics, safety considerations, responding agencies, and resource requirements.

	• It is imperative that all personnel operating at a search incident have appropriate personal protective equipment that corresponds with their assigned responsibilities and tasks and provides for their protection and safety. • Ensuring personnel accountability throughout these incidents is essential. • Communications system reliability, pre-search briefing, crowd control, heat exhaustion or heat stroke, incident safety officer (ISO), injuries, long-duration incidents, mental or psychological state of search subject, personal protective equipment, personnel accountability, rehabilitation, search area terrain and other hazards, stress management, traffic control, and weather (extreme temperatures, humidity, wind, precipitation)
Responding Agencies*	• Fire department, emergency medical services, law enforcement, specialized search teams, and support organizations
Resource Requirements	Personnel* • Fire department, search and rescue, emergency medical services, and law enforcement • In addition to human trackers, many search operations will utilize canines trained in trailing and tracking lost persons. Personal protective equipment* • Appropriate personal protective equipment will be based on the specifics of the incident and should correspond with assigned responsibilities and tasks during the search. Apparatus* • Firefighting apparatus, emergency medical units, rescue apparatus, specialized vehicles (all-terrain vehicles, mountain bikes, snowmobiles), aircraft (helicopters and fixed-wing), and support vehicles (communications, lighting, and rehabilitation units) Equipment* • The equipment required to effectively, efficiently, and safely conduct a search will be based on the specifics of the incident, as in the case of generators and fixed and portable lighting equipment when conducting a search at night. • Appropriate equipment will be required to establish and monitor the search area perimeter, as well as control crowds and vehicular traffic. • Reliable communications equipment, such as portable radios, will be needed at these incidents. • Some of the tools utilized in other rescue situations, such as thermal imaging cameras, may prove valuable in locating missing individuals based on the thermal signature created by a subject's body heat.

Fig. 18–4. Incident management job aid (continued)

	• Specialized equipment utilized at these incidents will provide capabilities in search area mapping, tracking missing persons, or personnel accountability. • *Information on additional equipment required in the rescue or recovery of lost persons is discussed in chapter 17.*

Fig. 18–4. Incident management job aid (continued)

Media Coverage: Searches	
WHO?	• Number of missing persons? • Personal information on missing person? • Age of missing person? • Skill and experience of missing person? • Subject's familiarity with involved area? • Was person alone or in company of others? • Medical condition of missing person? • Emotional or psychological state of missing person? • Clothing that missing person was wearing? • Equipment in possession of missing person? • Was missing person on foot? • Did missing person have a vehicle? • Missing person's ability to travel a distance? • Has missing person been found? • Has missing person been rescued? • Was missing person injured? • Did missing person experience a medical emergency? • Has body of missing person been recovered? • *Have appropriate notifications been made?* • *Has information on victims been received from a credible and authorized source?* • *Has information been verified for accuracy?* • How and to which medical facility was missing person transported? • Persons interviewed and relationships to missing person? • Witnesses who observed missing person? • Party who reported that person was missing? • *See chapter 21 for incidents involving civilian injuries or fatalities.*

Fig. 18–5. Media coverage job aid

WHAT?	• Type of search? • Conditions under which subject went missing? • Is this the first time subject has gone missing? • Where was person found in previous incident(s)? • What was person doing when he or she went missing? • Was activity planned or unplanned? • Responding agencies? • Resource requirements (number and types of personnel, apparatus, and equipment)? • Specialized resources utilized? • Incident priorities? • Operational tactics? • Safety considerations? • Incident management challenges? • Related situational and environmental factors? • Terrain of search area? • Hazards of search area? • Weather-related factors?
WHEN?	• Day of week? • Time of day? • Day/night? • Season of year? • Last time subject can be accounted for? • How long has person been missing? • Time search was initiated? • Time search was suspended? • Time search was resumed? • Time search was terminated? • Time search mode changed from rescue to recovery? • Time missing person walked out? • Time missing person was located? • Time missing person was rescued and removed to a safe location? • Time body of missing person was recovered?
WHERE?	• Location where missing person was last seen? • Direction in which person was headed? • Possible destination of missing person? • Municipality or neighborhood? • Size of search area? • Location of search area?

Fig. 18–5. Media coverage job aid (continued)

WHY?	• Factors that contributed to the subject going missing? • Was subject having personal problems? • Was subject having family issues? • Was subject under the influence of any substance? • Was subject suffering from a medical condition or illness? • Did subject have any psychological or emotional issues? • Was subject despondent? • *Has information on missing person been received from a credible and authorized source?* • *Has information been verified for accuracy?*
HOW?	• Was subject familiar with involved area? • Weather-related factors? • Terrain of involved area? • Situational and environmental factors?
Sidebar Story Ideas	• Alzheimer's • Canine searches • Human trackers • Managing search operations • Missing persons • Regional search and rescue capabilities • Roles in search operations • Search and rescue teams • Search equipment • Search techniques • Training search teams • Underground searches • Utilizing technology in search operations • Water searches • Wilderness searches
Resources	• Federal Emergency Management Agency (FEMA) www.fema.gov • National Association for Search and Rescue (NASAR) www.nasar.org • United States Fire Administration (USFA) www.usfa.fema.gov

Fig. 18–5. Media coverage job aid (continued)

Chapter Questions

1. Identify the typical categories of searches.
2. List the incident management priorities in managing searches.
3. Discuss several operational tactics related to searches.
4. Identify the safety concerns related to searches.
5. Identify the emergency service and support organizations and agencies that are dispatched to searches.
6. Discuss the resource requirements, in terms of personnel, apparatus, and equipment, associated with searches.
7. Relate and explain the essential story elements of news coverage of searches.

Chapter 19
Terrorism Incidents

Chapter Objectives

- Identify the typical categories of terrorism incidents.
- Identify the incident management priorities for terrorism incidents.
- Discuss the operational tactics utilized at terrorism incidents.
- Discuss the safety considerations associated with terrorism incidents.
- Identify the emergency service agencies that respond to terrorism incidents.
- Identify the resource requirements associated with terrorism incidents.
- Discuss the story elements associated with media coverage of terrorism incidents.

The threat of terrorism has become an ever-present reality in the contemporary world in which we live. The nature of this threat, as well as the associated potential consequences of such events, dictates that jurisdictions and the public safety, emergency service, and emergency management agencies that serve and protect them, be proactive in their attempts to prevent such events and in their preparedness to respond in the event that such an incident were to occur. The nature of these events dictates that law enforcement will take the lead in the management of these incidents, with fire and emergency services serving in appropriate support roles in mitigating the aftermath of the incident. Likewise, law enforcement agencies, including the Federal Bureau of Investigation (FBI) and the Bureau of Alcohol, Tobacco, Firearms, and Explosives (ATF), will be responsible for the investigation of the incident and the identification and criminal prosecution of all responsible parties (fig. 19–1).

Fig. 19–1. Law enforcement representative conducting a press conference. (Courtesy of Marple Township Police Department.)

The Federal Bureau of Investigation defines terrorism as "the unlawful use of force or violence against persons or property to intimidate or coerce a government, the civilian population, or any segment thereof, in furtherance of political or social objectives."[1] The FBI further categorizes terrorist acts as domestic or international, based on the origin, base, and objectives of terrorist groups. It should be noted that while terrorist acts are considered crimes, an investigation of a given event may reveal that while it involved criminal activity, it fell short of meeting the defined criteria for a terrorist event as defined above.

Since the 1990s the incidence of terrorist events in the United States and around the world has been increasing. While some of these events are not well known and received limited media coverage, events including the World Trade Center bombing in 1993, where six people were killed, and the 1995 bombing of the Alfred P. Murrah Federal Building in Oklahoma City, where 168 people were killed and hundreds injured, illustrate that as a nation we are vulnerable to the threat of terrorism. The tragic, unimaginable events that occurred during the September 11, 2001, attacks are definitive, dramatic illustrations that terrorism is an ever-present reality in the contemporary world in which we live, work, and travel.

Typical Incidents

Terrorist incidents can be classified into five types of events, as listed in figure 19–2. While many terrorism incidents will involve only a single type, the use of multiple types of terrorism against a target should not be ruled out. Likewise, the simultaneous deployment of more than one terrorist act at a single or multiple locations is possible. The five recognized categories of terrorist acts are biological events, chemical events, explosive events, incendiary events, and nuclear events. *Biological events* involve the release of biological agents such as anthrax, bacteria, toxins, or viruses. Typical agents used in *chemical events* include blister agents, which cause severe burns; blood agents, which interfere with the transmission of oxygen through the bloodstream; choking agents, which can cause severe respiratory distress; irritating agents, which cause problems with the respiratory system and the skin; and nerve agents, which disrupt the transmission of nerve impulses within the body.

Potential threats	*Potential harm*	*General targets*
• Biological	• Asphyxiation	• Controversial businesses
• Chemical	• Chemical	• Historical sites
• Explosive	• Etiological	• Infrastructure systems
• Incendiary	• Mechanical	• Places of assembly
• Nuclear	• Psychological	• Public buildings
	• Radiological	• Symbolic targets
	• Thermal	

Fig. 19–2. Typical incidents: terrorism incidents

The majority of terrorist attacks around the world have involved the use of explosives. These *explosive events* involve the use of substances and materials designed to rapidly generate and release the byproducts of an explosion. With all such events, the potential of a secondary device or devices is a primary concern. *Incendiary events* involve the use of chemical, electrical, or mechanical energy to start a fire. Further information on incendiary fires can be found in chapter 23. *Nuclear events* constitute the final, and perhaps ultimate, type of terrorist act. The potential for widespread harm and destruction associated with nuclear devices, as well as the proliferation of nuclear weapons and capabilities, represent a growing concern in the unstable and unpredictable contemporary world.

In addition to understanding the potential threats of domestic and international terrorism, those public servants responsible for protecting the communities that they serve must fully understand and be prepared to address the potential harm that can result from terrorist acts. Ensuring life safety will obviously take precedence in all actions to protect the public before the occurrence of an event or to mitigate the consequences of such an event. The harm that can be inflicted by a terrorist act can involves *asphyxiation* that deprives the human body of necessary oxygen, *chemical* burns resulting from exposure to corrosive or toxic materials, *etiological* harm resulting from exposure to disease-causing organisms or toxins derived from living organisms, *mechanical* harm from physical trauma, *psychological* harm from the mental stress and anguish associated with experiencing such events, *radiological* harm from exposure to high levels of radiation, or *thermal* harm resulting from exposure to extremely hot or cold temperatures.

While the specific targets of a terrorist or terrorist group will usually align with their intended purpose in conducting an attack, often including harming both the public and emergency response personnel, there are certain general targets that attract the attention of terrorists. These attacks are thus often referred to as involving weapons of mass destruction (WMD). General targets of terrorist attacks include controversial businesses, historical sites, infrastructure systems, places of assembly, public buildings, and symbolic targets. Analysis of the targets of the tragic events of September 11, 2001, reveals that the World Trade Center, whose towers were targeted by terrorists piloting two commercial airplanes carrying passengers and a massive quantity of jet fuel, involved several of these recognized general targets.

Incident Priorities

As with all incidents, the priorities in the management of terrorism incidents, in priority order, are (1) life safety, (2) incident stabilization, and (3) property conservation.

Operational Overview

While law enforcement agencies, fire departments, and other emergency service organizations engage in extensive planning in the event that a terrorist event were to occur within their community or within a neighboring jurisdiction where they may be called upon to provide assistance through mutual aid, there will typically be many challenges associated with the effective, efficient, and safe management of these incidents. These challenges often begin at the time of the initial dispatch of emergency response units and personnel. There will be instances where, as a result of credible threats received in advance of the incident or based on the information provided by reporting parties, the incident is determined to involve a potential act of terrorism and is appropriately dispatched as such, alerting emergency responders to the nature of the incident and the risks, hazards, challenges that may be inherent in its management. Unfortunately, there will likewise be instances where such information does not become available before or subsequent to the initial dispatch and is only discovered after the arrival of emergency responders on an incident scene. The crucial importance of response personnel from all disciplines having the necessary training, knowledge, and skills to determine in a timely manner that an incident may be the result of an act of terrorism and acting accordingly in terms of managing the incident and requesting appropriate resources and assistance is obvious. As always, the priority of ensuring the life safety of emergency responders and the public will guide all actions and decisions. The nature of these incidents demands that in addition to conducting a comprehensive initial situational assessment, ongoing assessments must be performed throughout the duration of the incident.

While the specific nature of each incident will influence the appropriate incident management approach in terms of strategic goals, operational mode, and tactics, and exhaustive coverage of the management of the various types of terrorist events is beyond the scope of this book, a number of general practices in the management of these incidents are considered. First and foremost is that an identification that an incident either has or may have resulted from an act of terrorism must occur as early as possible in the management of the incident. In cases where that information is available while emergency response units are responding to the incident, it will trigger the cautious staging of all apparatus and personnel an appropriate distance from the incident. Situational awareness on the part of emergency responders will enable them to observe and consider indicators that the incident to which they have responded may, in fact, be the result of an act of terrorism. Were the fire department to respond to a reported building fire involving an occupancy that had been the subject of recent credible threats, represented a general or specific terrorist target, or involved a fire that started under seemingly unusual circumstances, it is imperative that the potential of terrorism be recognized and guide all subsequent actions and decisions.

Timely scene assessment and management should result in the establishment of appropriate control zones and perimeter controls in the interest of protecting both the public and emergency responders. In those instances where it is only subsequently determined that an incident may be the result of a terrorist act, immediate actions must be taken to move all involved personnel to an area that provides for their safety. This determination should be communicated in a timely manner to the jurisdiction's communications center, which will initiate the timely dispatch of additional resources and make appropriate notifications to all agencies having jurisdiction. While law enforcement personnel will assume a lead role in the management of these incidents, the decision will often be made to utilize a unified command structure in the interest of enhancing the effectiveness, efficiency, and safety of incident management.

These incidents, like hazardous materials incidents, demand that responders refrain from proactive action until such time as a comprehensive situational assessment and incident surveillance has resulted in a determination that emergency responders can be safety deployed to enact their normal duties, as in the case of extinguishing a fire, or searching for and rescuing victims. The nature of many of these incidents will be such that they may result in a number of victims, as in the case of an incendiary fire in a building, an explosive device causing a building collapse that traps and injures victims, or the dissemination of a biological agent in the air handling system of a building or other structure, such as an underground train or subway station. The potential of a secondary device(s) designed to activate after emergency response personnel have entered the involved area to address the initial incident must always be anticipated and addressed.

In addition to establishing perimeters in the interest of crowd and traffic control, providing for scene security will be essential at any suspected terrorist incident. Denying entry to the incident scene, which in the case of a terrorist act would also be a crime scene, as well as ensuring that essential evidence is preserved, properly collected by trained investigators, and that the required chain of custody is maintained in the handling of all evidence collected, represent critical elements in the successful investigation and resulting prosecution of responsible parties.

The nature of these incidents often produces casualties in terms of injuries and fatalities. Once it has been determined that emergency responders can assume operating positions on the incident scene, search and rescue operations can be undertaken. As victims are located, emergency medical services personnel will provide needed patient care, assessing and possibly triaging victims before their subsequent treatment and transport to appropriate medical care facilities. These incidents, particularly those that involve numerous victims, are often capable of overwhelming a community's healthcare system. It should be noted that most healthcare organizations have, likewise, been engaged in the related planning and preparedness activities discussed earlier in this chapter.

Safety Considerations

While most incidents to which emergency responders are dispatched involve risks and hazards that are fairly well understood and addressed in incident management, as in the case of a building that is on fire losing its structural integrity and subjecting firefighters to potential harm were the building to experience a floor, wall, or roof collapse, the potential inherent risks of operating in the aftermath of a terrorist attack can expose emergency responders to extremely dangerous situations and operating environments. As with all other emergency incidents, life safety must guide all decisions while managing the incident and actions while operating on the incident scene. In addition to sealing off the involved area to people and traffic to ensure their safety and maintain the security of the scene that may be determined to be a crime scene, control zones should be established and enforced until such time as the involved hazards and threats have been evaluated and a determination has been made by law enforcement and appropriate technical specialists that it is safe for emergency responders to be deployed to perform their assigned tasks, such as fire suppression or search and rescue.

Only personnel with the appropriate training, qualifications, and credentialing should be permitted to operate at these incident scenes and, as always, should wear appropriate personal protective equipment consistent with their given responsibilities and assignments. The importance of incident command, control, and coordination, as discussed in chapter 8, bears repeating with respect to these incidents, as does the necessity of an appropriate incident management system, the utilization of an incident safety officer (ISO), and tracking of personnel accountability throughout these incidents. Having all emergency response personnel operating at these incidents working together and clear on the roles and responsibilities is imperative and is often achieved through the establishment of a unified command.

These incidents can present many potential threats capable of invoking serious harm to emergency responders, including injuries, illnesses, and fatalities. The potential routes of exposure discussed earlier—absorption, ingestion, and inhalation—must be addressed through the use of appropriate personal protective equipment, including respiratory protection. Technological innovation has provided many tools that contribute to the enhanced safety of emergency responders at these incidents, including bomb robots that can stand in for responders during high-risk operations. The ever-present threat of secondary devices makes the use of such tools a critical aspect of ensuring the life safety of emergency personnel operating at these incidents.

Resource Requirements

In addition to the local fire, emergency medical services, law enforcement, and technical resources that routinely respond to incidents within a community, these incidents will necessitate the response of an expanded cadre of additional local, regional, and, at times, federal resources. Local emergency management agencies will have an integral role at these incidents, often being assisted by their state counterparts or the Federal Emergency Management Agency (FEMA) and other resources from the Department of Homeland Security (DHS). The nature of a particular incident will trigger the involvement of such resources as a bomb squad. Appropriate investigative agencies, including the FBI and the Bureau of Alcohol, Tobacco, Firearms, and Explosives (ATF), will often be involved in the investigation of these incidents and subsequent prosecution of responsible parties. The overall resource needs will obviously be determined by the nature and specifics of a given incident, with some of these incidents necessitating the involvement of public works agencies as well as public utilities. Given the large number of resources that may be involved in the management of these incidents, including both assisting and cooperating agencies, the use of an appropriate ICS organizational structure, including the utilization of the command staff position of liaison officer, is often both necessary and beneficial.

The apparatus and equipment needs of these incidents will once again correspond to the nature and scope of an incident and its resulting resource needs. Appropriate apparatus and equipment will be deployed to the incident from each of the above organizations and agencies in accordance with the resource needs of the incident and the operating assignments that each organization and its personnel will be required to perform in the interest of effectively, efficiently, and safely resolving the incident. Further information on necessary personnel, personal protective equipment, apparatus, and equipment can be found in the earlier incident-specific chapters.

Media Coverage of Terrorism Incidents

Incidents that result from acts or suspected acts of terrorism will without question be news stories in which all news media organizations and the audiences that they serve will have an interest. The interest in, as well as news coverage of, these incidents will almost instantaneously spread across the nation and around the world. The advent of the Internet and social media, as discussed in chapter 7, will often be a contributing factor in alerting the public to the occurrence of these incidents and prompting their interest in receiving not only initial information, but also updates of all significant developments throughout the duration of the incident.

The magnitude of the challenges of not only managing these incidents, but ensuring effective media coverage, will often intensify as these breaking news stories instantaneously advance to the national and frequently international stage as news organizations from across the nation and the world initiate coverage within their 24/7 news cycles. As with all incidents, these incidents begin as local events, but can quickly result in the deployment of additional regional, state, and at times, federal resources. Likewise, news organizations from across the nation, as well as around the world, may incorporate local news coverage into their broadcasts, and often pick up incident feeds that depict real-time images from the incident scene. Given the newsworthiness of these stories, these news organizations may just as frequently dispatch their reporters,and photojournalists, along with the tools of their trade, to make the pilgrimage to the incident scene to provide on-scene incident reporting. The many challenges, including logistics, of hosting such an invasion by the news media during incidents of national consequence or interest, as well as suggestions on how to ensure effective media coverage at these high profile and visibility incidents, is considered in chapter 24.

While these are incidents that by their very nature and significance will trigger a desire on the part of the public to receive information in a timely manner, the other stakeholder expectations that coverage also be accurate, comprehensive, and professional are also imperative in coverage of these incidents. These are stories where inaccurate, confusing, or misleading information can produce a variety of detrimental outcomes. All information incorporated in related news reporting must be verified for accuracy and credibility. The sources from which information is received must be credible and trustworthy. It must be recognized that during the early stages of these incidents, those with incident management responsibilities will without question need to engage in assessing the situation and developing and implementing the necessary tactical operations required to effectively, efficiently, and safely resolve the incident.

During this crucial time, during which the early decisions made will often influence the success of the overall incident outcome, it is imperative that news media representatives understand and respect the fact that the incident commander and others may not be available to speak with the media. The utilization of a qualified PIO at these incidents will be instrumental in ensuring that the early incident coverage is informed and accurate. It must also be recognized that, given that these incidents will often involve criminal activity and an active criminal investigation targeted at identifying and prosecuting responsible parties, certain information must remain confidential in the interest of not compromising the successful investigation and criminal prosecution of the responsible individuals or groups. Through the cooperative working relationships discussed throughout the early chapters of this book, emergency personnel and their media counterparts can ensure that all media coverage of these tragic events is accurate, comprehensive, professional, and timely. As with many other types of incidents, public safety agencies may seek the assistance of the media in the dissemination of incident-related information to the public or to request the assistance of individuals who may have observed suspicious activity or have knowledge of events related to the incident.

The job aids in figures 19–3 and 19–4 are designed to serve as resources for fire department and media personnel. Figure 19–3 provides an overview of incident management of terrorism incidents, while figure 19–4 provides guidance with respect to media coverage of these incidents.

Incident Management: Terrorism Incidents	
Typical Incidents	• Biological events • Chemical events • Explosive events • Incendiary events • Nuclear events • Potential harm (asphyxiation, chemical, etiological, mechanical, psychological, radiological, thermal) • General targets (controversial businesses, historical sites, infrastructure systems, places of assembly, public buildings, symbolic targets)
Incident Priorities	1. Life safety 2. Incident stabilization 3. Property conservation
Operational Tactics*	• Air monitoring, assess hazards, assess threats, control zones, crowd control, decontamination, establish crime scene, evacuation, exposure protection, extinguishment, fire attack, investigation, patient care (assessment, triage, treatment, transport), perimeter control, personnel accountability, post-incident analysis, preserve evidence, recovery, rehabilitation, rescue, sample collection and analysis, scene assessment, scene management, scene security, search, select appropriate operational mode, stabilization, staging, technical rescue, traffic control, and utility control • Appropriate operational tactics should be developed based on the nature and dynamics of the incident. • The operational tactics discussed in the earlier incident-specific chapters will be appropriate when responders are deployed to perform those types of assignments, as in the case of firefighting, hazardous materials management, or search and rescue operations. • Personnel should not be given assignments until it has been determined that the scene is both secure and safe for them to operate.
Safety Considerations*	• These incidents are known to involve inherent risks that can expose emergency responders to extremely dangerous situations and operating environments. • The priority of life safety must guide all decisions and actions while managing and operating at these incidents.

Fig. 19–3. Incident management job aid. **Note:* The nature of a particular incident will determine the appropriate operational tactics, safety considerations, responding agencies, and resource requirements.

	• Emergency responders should not be deployed into hazard areas until such time as these areas can be evaluated in terms of hazards and/or threats to life safety. • Until such time as law enforcement or technical specialists can assure a safe working area, fire department personnel should stage in readiness rather than make entry into an area of unknown hazards and risks. • An ever-present concern at these incidents will be the potential of secondary devices. • All personnel operating at these incidents must have the appropriate training and qualifications that correspond with the tactical assignments that they are given. • All personnel operating at these incidents must utilize the appropriate level and type of personal protective equipment based on their work assignments. • Personnel accountability must be ensured at all times while operating at these incidents. • Breathing air supply, building collapse, confined spaces (limited access, limited ventilation, potentially harmful atmospheres), control zones, critical incident stress debriefing, crowd control, contamination/decontamination, evacuation, explosions, explosive atmospheres, exposure (absorption, ingestion, inhalation), hazards, hazardous materials, IDLH atmospheres, ignition sources, illnesses, incendiary devices, incident safety officer (ICS), Mayday (becoming trapped, disoriented, lost, or running out of breathing air), perimeter control, personnel accountability, post-incident analysis, rapid intervention teams, rehabilitation, scene assessment, scene management, scene security, secondary devices, stabilization, toxic atmospheres, traumatic injuries, utilities, vehicular traffic, and victims
Responding Agencies*	• Fire department, law enforcement, emergency medical services, bomb squads, hazardous materials teams, technical and specialized rescue teams, emergency management agencies, fire investigation units, criminal investigation units, public utilities (electric, gas, sewer, telephone, water), and public works • Law enforcement will usually serve as the lead agency at these incidents. • The nature of a particular incident will determine the appropriate agencies that will be required in the management, investigation, and prosecution of the incident. • Local resources will often be assisted by other agencies that have jurisdiction, such as the Federal Bureau of Investigation (FBI), or regional or federal resources, such as search and rescue teams.

Fig. 19–3. Incident management job aid (continued)

<table>
<tr><td>Resource Requirements</td><td>Personnel*
• Fire department, law enforcement, emergency medical services, hazardous materials teams, technical and specialized rescue teams, emergency management, fire and criminal investigators, public utilities, and public works
Personal protective equipment*
• All personnel operating at these incidents should wear appropriate personal protective equipment based on their assigned responsibilities and tasks on the incident scene.
Apparatus*
• Firefighting apparatus, rescue apparatus, aerial apparatus, water supply apparatus, emergency medical units, specialized apparatus, support apparatus, public works and heavy construction vehicles, and utility company vehicles
Equipment*
• The equipment needs of these incidents will be determined by the nature of a particular incident and the resulting assignments that personnel will be deployed to complete.
• Emergency responders involved in firefighting, hazardous materials management, search, and/or rescue activities will utilize the equipment discussed in the earlier incident-specific chapters.
• Technical and specialized rescue teams will utilize specialized tools and equipment, including advanced technology search equipment.
• Personnel with law enforcement and investigative responsibilities will utilize the specific tools of their profession, including specialized equipment related to the needs of the particular incident.</td></tr>
</table>

Fig. 19–3. Incident management job aid (continued)

Media Coverage: Terrorism Incidents	
WHO?	• Number of victims? • What were victims doing when the event occurred? • Number of and extent of civilian injuries? • Number of and cause of civilian fatalities? • Number of and extent of emergency responder injuries? • Number of and cause of emergency responder fatalities? • Number of missing or unaccounted for persons? • Personal information on injured individuals?

Fig. 19–4. Media coverage job aid

	• Personal information on fatalities? • Personal information on missing or unaccounted for persons? • *Have appropriate notifications been made?* • *Has information on victims been received from a credible and authorized source?* • *Has information been verified for accuracy?* • How and to which medical facilities were victims transported? • Medical condition of victims (initial and present)? • *See chapter 21 for incidents involving civilian injuries or fatalities.* • *See chapter 22 for incidents involving firefighter injuries or fatalities.* • Responsible individual(s) or group(s)? • Individual(s) sought for questioning? • Individual(s) taken into custody? • Individual(s) charges with a crime? • *Has information on responsible parties been received from a credible and authorized source?* • *Has information on responsible parties been verified for accuracy?*
WHAT?	• Type of incident? • Domestic terrorism? • International terrorism? • Biological event? • Chemical event? • Explosive event? • Incendiary event? • Nuclear event? • Secondary device? • Resulting harm? • Hazardous material(s) involved? • Properties of involved hazardous materials? • Buildings involved? • Vehicles involved? • Other property involved? • Exposed buildings? • Other exposed property? • Extent of damage? • Damage to other buildings or property? • Damage to vehicles? • Responding agencies? • Assisting agencies? • Cooperating agencies?

Fig. 19–4. Media coverage job aid (continued)

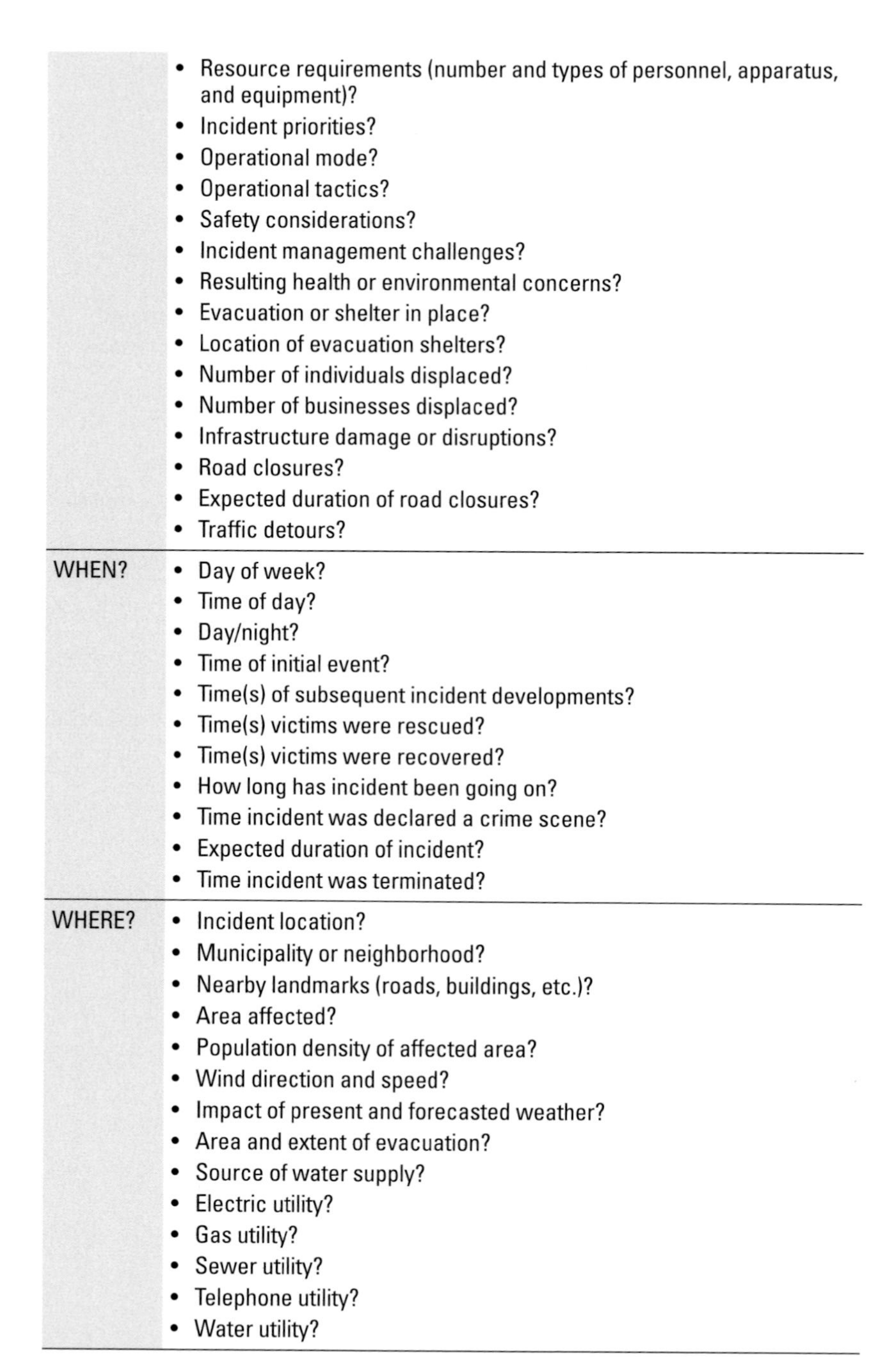

	• Resource requirements (number and types of personnel, apparatus, and equipment)? • Incident priorities? • Operational mode? • Operational tactics? • Safety considerations? • Incident management challenges? • Resulting health or environmental concerns? • Evacuation or shelter in place? • Location of evacuation shelters? • Number of individuals displaced? • Number of businesses displaced? • Infrastructure damage or disruptions? • Road closures? • Expected duration of road closures? • Traffic detours?
WHEN?	• Day of week? • Time of day? • Day/night? • Time of initial event? • Time(s) of subsequent incident developments? • Time(s) victims were rescued? • Time(s) victims were recovered? • How long has incident been going on? • Time incident was declared a crime scene? • Expected duration of incident? • Time incident was terminated?
WHERE?	• Incident location? • Municipality or neighborhood? • Nearby landmarks (roads, buildings, etc.)? • Area affected? • Population density of affected area? • Wind direction and speed? • Impact of present and forecasted weather? • Area and extent of evacuation? • Source of water supply? • Electric utility? • Gas utility? • Sewer utility? • Telephone utility? • Water utility?

Fig. 19–4. Media coverage job aid (continued)

WHY?	• Responsible individual(s) or group(s)? • Political objective(s)? • Social objective(s)? • Motives behind terrorist act? • Desire to make a political statement? • Desire to draw attention to a cause? • Desire to harm civilians? • Desire to harm emergency responders? • Desire to disrupt economy or financial system? • Desire to disrupt infrastructure? • Desire to diminish public sense of safety and security? • Desire to disrupt government or public confidence in it?
HOW?	• Biological agent(s)? • Chemicals involved? • Explosion? • Incendiary device? • Nuclear device? • Secondary device(s)? • Official determination of incident cause(s)? • Is the incident being investigated? • Agencies involved in incident investigation? • Are charges being brought against responsible parties? • Criminal charges?
Sidebar Story Ideas	• Aviation security • Biological events • Border security • Chemical events • Chemical security • Cybersecurity • Department of Homeland Security • Domestic terrorism • Explosive events • Incendiary events • Information sharing (fusion centers) • International terrorism • National security • Nuclear events • Nuclear security • Regional terrorism task forces • Reporting suspicious activities

Fig. 19–4. Media coverage job aid (continued)

	• Roles of governmental agencies • Roles of emergency response organizations • Targets of terrorism • Terrorism preparedness • Terrorism recovery • Terrorism response • Vulnerability assessment • Warning systems
Resources	• Bureau of Alcohol, Tobacco, Firearms, and Explosives (ATF) www.atf.gov • Centers for Disease Control and Prevention (CDC) www.cdc.gov • Federal Bureau of Investigation (FBI) www.fbi.gov • Federal Emergency Management Agency (FEMA) www.fema.gov • U.S. Department of Homeland Security (DHS) www.dhs.gov • United States Fire Administration (USFA) www.usfa.fema.gov

Fig. 19–4. Media coverage job aid (continued)

Chapter Questions

1. Identify the typical categories of terrorism incidents.
2. List the incident management priorities in managing terrorism incidents.
3. Discuss several operational tactics related to terrorism incidents.
4. Identify the safety concerns related to terrorism incidents.
5. Identify the emergency service and support organizations and agencies that are dispatched to terrorism incidents.
6. Discuss the resource requirements, in terms of personnel, apparatus, and equipment, associated with terrorism incidents.
7. Relate and explain the essential story elements of news coverage of terrorism incidents.

Notes

1 Federal Bureau of Investigation. 2005. *Terrorism 2002–2005*. Reports and Publications. http://www.fbi.gov/stats-services/publications/terrorism-2002-2005.

Chapter 20

Weather-Related Events and Natural Emergencies

Chapter Objectives

- Identify the typical categories of weather-related events and natural emergencies.
- Identify the incident management priorities for weather-related events and natural emergencies.
- Discuss the operational tactics utilized at weather-related events and natural emergencies.
- Discuss the safety considerations associated with weather-related events and natural emergencies.
- Identify the emergency service agencies that respond to weather-related events and natural emergencies.
- Identify the resource requirements associated with weather-related events and natural emergencies.
- Discuss the story elements associated with media coverage of weather-related events and natural emergencies.

Some of the incidents to which fire departments are dispatched are the result of severe weather or natural disasters. While the nature of these incidents will vary in accordance with the particular emergency situation, the fire department is typically dispatched to address the situation resulting from the event or its aftermath through ensuring life safety and, when possible, saving involved or threatened property. As always, life safety will take precedence over property conservation.

These incidents often challenge the resources and capabilities of a community, including those of fire and emergency service organizations, based on the nature, size, scope, or duration of the event. While these incidents often trigger the assistance of mutual and outside aid, frequently these events impact entire regions and even states rather than an isolated community, thus often significantly reducing available resources to assist local communities and the emergency service organizations that serve and protect them. While the fire department will have the primary responsibility or take a lead role in the management of many of the types of emergency incidents discussed in the earlier chapters of this book, weather-related events and natural disasters often fall within the jurisdiction of emergency management agencies, with the fire department serving in a support capacity. This role will vary in accordance with the specific needs of the incident and the resources and capabilities of the fire department and could involve rescuing trapped or stranded individuals, assisting with evacuations, pumping out flooded areas or structures, or extinguishing a fire that started as a result of a lightning strike during a thunderstorm.

Typical Incidents

While severe weather and natural disasters can impact any community, the vulnerability of a given community to these events will often be in large part based on its geographic location. For example, communities in nearby proximity to an active volcano will face the potential for periodic eruptions, while most other communities would face minimal risk of being affected by this event. Likewise, particular geographic areas are more prone to certain weather conditions such as hurricanes, tornadoes, cyclones, typhoons, or tsunamis.

Severe weather and natural disasters fall into an array of events or incidents as listed in figure 20–1. The frequency with which some of these events may impact a given region or community is usually difficult to predict. Earthquakes are illustrative of this reality. Other severe weather events, while typically impacting particular geographic regions more than others, have the potential of affecting a great majority of communities, as in the case of thunderstorms and lightning. Many severe weather events have the potential of leaving a trail of destruction, including damage from flooding and high winds (fig. 20–2). In certain areas heavy rains and flooding can produce mudslides, which in addition to threatening lives, can also damage or destroy structures and community infrastructure including highways and utilities.

- Cyclone
- Drought
- Earthquake
- Extreme temperature
- Flood
- High winds
- Hurricane
- Ice storm
- Lightning
- Mudslide
- Natural disaster
- Severe weather
- Snowstorm
- Thunderstorm
- Tornado
- Tsunami
- Typhoon
- Volcanic eruption
- Winter storm

Fig. 20–1. Typical incidents: weather-related events and natural emergencies

Fig. 20–2. Flooding during a hurricane

The season of the year also influences the occurrence of severe weather events. For example, extreme temperatures and lack of humidity during the summer months can set the stage for the occurrence of wildland fires that can ravage numerous acres of land and the structures that reside on it, as discussed in chapter 13 (fig. 20–3). The same severe weather conditions that contributed to these fires likewise present numerous life safety and operational challenges to the firefighters responsible for confining and extinguishing these often massive and long-duration fires. Many geographic areas are similarly vulnerable to winter storms involving snow, ice, or blizzard conditions, which frequently contribute to accidents, as discussed in chapter 9.

Fig. 20–3. Posted fire warning during wildland fire season. (Courtesy of Joe Landy.)

Incident Priorities

As with all incidents, the priorities in the management of a severe weather event or natural disaster, in priority order, are (1) life safety, (2) incident stabilization, and (3) property conservation.

Operational Overview

The role of the fire department at these incidents will vary based on the nature and consequences of these incidents, with the fire department typically contributing to consequence management. The aftermath of certain events will yield emergencies that will require the fire department to deliver services that it routinely performs in times of isolated incidents such as building fires, hazardous materials incidents, and infrastructure emergencies. Many of these incidents will involve the fire department engaging in search and rescue operations in the interest of locating victims and ensuring their safety. Frequently, the fire department will participate in the evacuation of a community and assist in relocating individuals from harm's way to the safety provided by an evacuation shelter. There will be times that the fire department may be called upon to utilize the drafting capabilities of its fire apparatus and portable pumps to relieve flooding in buildings and low-lying areas, including those housing critical community infrastructure, such as a hospital or power plant.

While many of the above tasks may clearly fall within the realm of services routinely provided by fire departments, the scope of the service needs during a major severe weather event or natural disaster differentiate these responses from the norm. The increased life safety hazards associated with many of these incidents, such as downed power lines in flooded areas, further challenge emergency responders at these incidents, as does the extensive and widespread destruction that these events can yield, often including the homes of emergency response personnel. The fact that some weather events can last for days and emergency responders are often required to engage in operations while the weather event is ongoing likewise present numerous challenges, particularly with respect to ensuring the life safety of response personnel.

As always, the effective, efficient, and safe management of these incidents begins with a comprehensive size-up. Unlike the typical size-up conducted at an incident such as a building fire where the incident commander walks around the outside of the building to conduct a 360-degree evaluation, oftentimes the fire department will be called upon to conduct a "windshield damage assessment" as crews are deployed to drive through the community in fire apparatus in the interest of detecting emergency situations requiring immediate or urgent attention, as well as to begin the damage assessment process that will be required if the community is to receive federal disaster assistance. Unfortunately, in many instances such a driving survey will not be possible as a result of roads that are flooded, blocked by debris, or unsafe to travel as a result of downed electrical power lines or broken gas mains. The nature and scope of these incidents may trigger the decision to utilize a unified, rather than a single command, as discussed in chapter 8.

Specific guidance on handling the various emergency situations resulting from severe weather and natural disasters is providing in these chapters:

- Chapter 9 Accidents
- Chapter 10 Building Fires
- Chapter 11 Equipment Fires
- Chapter 12 Hazardous Materials Fires
- Chapter 13 Outside Fires
- Chapter 14 Vehicle Fires
- Chapter 15 Hazardous Materials Incidents
- Chapter 16 Infrastructure Emergencies
- Chapter 17 Rescues
- Chapter 18 Searches

The incident management job aids in each of these chapters provide specific information regarding operational tactics, safety considerations, and resource requirements at each type of incident.

Safety Considerations

While the appropriate safety considerations will vary based on the specific nature of these incidents, each type of resulting emergency incident presents its own inherent challenges in terms of ensuring the life safety of both the public and emergency responders. Pertinent related information can be found in the preceding incident-specific chapters. As always, the incident management priority of life safety must drive all decisions and actions on the incident scene. The importance of thorough planning and coordination of all actions, rather than allowing freelancing to take place, must be emphasized. This will require the utilization of an incident command system, an incident safety officer (ISO), and a personnel accountability system. At times, rather than utilizing a single command, a unified command will be established in the interest of effective, efficient, and safe incident operations and ensuring appropriate coordination and control.

The provision of reliable communication systems will be imperative at these incidents, with interoperability of radio communications being extremely important when different response disciplines and emergency responders from other organizations or areas are working together at these incidents. It is crucial that firefighters and other personnel have the ability at all times to report essential information regarding their progress in conducting searches and rescues, hazardous conditions that they discover, and other relevant information required for the successful management of the incident. Reliable radio communications are extremely important should firefighters find themselves trapped, disoriented, lost, injured, or in harm's way, as in the case of raging waters or a building collapse.

It is imperative that all emergency responders only operate to the level to which they are trained and equipped, calling for specialized teams such as hazardous materials or technical rescue teams when needed. It is likewise imperative that appropriate personal protective equipment (PPE) be available and utilized in accordance with the nature and needs of the specific emergency situation. While firefighters engaged in battling a building fire that started after a lightning strike will require the protection of structural turnout gear, that same gear would compromise the safety and health of a wildland firefighter fighting a fire originating from that same weather event. Personnel operating in environments where hazardous materials are present must wear the appropriate level of protection based on the specific materials involved. Likewise, rescue personnel will utilize appropriate PPE based on their assigned tasks, as in the case of the use of personal flotation devices for those engaged in water rescues.

While the nature of the incidents that arise as a result of weather events and natural disasters can vary widely as discussed above, it is not unusual for severe weather or a natural disaster to produce more than one of the types of emergency incidents discussed in earlier chapters. A tornado or earthquake, for example, could result in accidents, various types of fires, hazardous materials incidents, infrastructure emergencies, and incidents involving search and rescue. Both rising and moving water may present significant hazards at some of these incidents. As always, the issues associated with hazardous materials involvement must be recognized, as should the dangers associated with the interaction of electricity and water, gas and water, or gas and electricity.

The safety concerns associated with many of these incidents increase in situations where an entire community or region has been impacted by the event. When this is the case, safety must be considered not only while operating on a specific incident scene, but also while traveling to and from each incident. Downed power lines, trees, and debris can make such travel more difficult or at times impossible. It must be recognized that severe weather and natural disasters can instantaneously wreak damage or destruction on a community, some of which may significantly increase the inherent hazards and potential for harm to both the public and emergency response personnel. Apparatus and personnel staging and the use of control zones are extremely important at many of these incidents, as is the use of perimeter and traffic controls to ensure the protection of both groups. The nature of the incident may also require life safety actions such as sheltering in place or evacuations.

Many severe weather events serve to remind us of nature's power and how vulnerable a community, its residents, and the public safety personnel who protect it can be at times. The fact that we are often at the mercy of the weather is even more of an issue should the weather event be of a protracted duration that spans many hours or even a number of days. This illustrates the importance of continuing to monitor current, approaching, and forecasted weather throughout the management of these incidents. While the original weather event will end at some point, its impact or aftermath, such as flooding, can continue for some time during which emergency operations, including rescues, continue to be conducted, and at times for an even longer period during which the community embarks on the pilgrimage of recovering from the tragedy. Thus while the storm may end, the storm clouds pass, and the skies turn sunny, the challenging work of emergency responders may continue for days, weeks, or months.

Resource Requirements

The typical response to these incidents will obviously vary based on the specific nature of the incident, as well as its scope and resource requirements. In addition to the typical response, including the fire department, emergency medical services, law enforcement, and appropriate specialized teams with capabilities in such areas as technical rescue or hazardous materials, public works and utility company personnel will frequently be required. Local emergency management agencies will typically play a lead role in the overall management of many of these incidents, with assistance from their state and federal counterparts as necessary. These are incidents where the use of the National Incident Management System (NIMS) is essential. The apparatus and equipment necessary to resolve these incidents will likewise be determined by the nature and resource needs of a specific incident, as discussed in the earlier incident-specific chapters.

Media Coverage of Weather-Related Events and Natural Emergencies

The geographic footprint of severe weather events and natural disasters can vary from a local event such as a tornado that strikes only one community, to a more powerful and expansive tornado that invokes its damage and destruction over a larger geographic region often involving more than one state, to an earthquake, hurricane, or winter storm that impacts an entire region of the nation. While all of these incidents will certainly be of interest to the residents of the impacted areas, many of them will also be of regional, national, and at times international interest. The level of interest of a news organization's readers, listeners, or viewers will usually have a significant influence on its determination of the newsworthiness of the story and related decisions on the resources to devote to its coverage. It is typical to deploy both reporters and photojournalists to provide coverage of many of these incidents. Ground reporting as well as aerial coverage affords a news organization the opportunity to provide both the big picture, as well as the specifics of the incident, including human interest stories. In those instances where regional, national, or international interest develops with respect to covering the story, these news organizations may send their own personnel to the incident scene or incorporate the coverage provided by local reporters and photojournalists into their news coverage. Incidents of national consequence or interest are considered in chapter 24. By their very nature, these incidents are often stories that illustrate the value of incorporation of visual images into news reporting. In the event of a hurricane, the inclusion of real-time visual images can be used to depict the storm's strength and the subsequent damage and destruction left in its aftermath. Accompanying visual images generated to show the past, current, and forecasted weather through the use of weather maps enhance the effectiveness of this news coverage, as do graphical images showing such critical factors as rainfall totals and flood stages.

While residents of affected areas will routinely look to news radio and television stations for current, up-to-the-minute news reporting on these events, there will be times that although these news organizations are prepared to stay on the air using power supplied by their emergency backup generators, members of their listening or viewing audience may not be as fortunate after losing their electrical power. The potential of damage to or destruction of a community's or region's infrastructure highlights the merit of making provision to communicate to the public through the various forms of communication and technology discussed in chapter 7, including the Internet and social media. While the power to a resident's home may be out, he or she may still have the capability of receiving pertinent information through battery-powered devices such as radios and smart phones.

These incidents further illustrate the important shared responsibility that emergency management officials, fire and emergency service organizations, and the news media have with respect to ensuring the life safety of those who live in, work in, or travel to or through a community. Through the cooperation discussed earlier, these organizations can collaborate to ensure that stakeholders receive situational updates, warnings, alerts, and instructions in an accurate, comprehensive, professional, and timely manner that minimizes confusion and misleading information, some of which can be originated and propagated by the many individuals who will desire to discuss the event on the Internet or through social media. The potential for inaccurate, confusing, or misleading information during these dangerous events can yield tragic consequences, as well as complicate the work of emergency responders if the recipients of such information view it as credible and unwisely act on it, such as deciding to shelter in place during a hurricane after a mandatory evacuation has been declared or to go out on the roadways and attempt to navigate through swift moving water only to discover that it often only takes a foot of water to capture and float a car.

This shared role of information dissemination and media coverage in severe weather events and natural disasters should be embraced and supported by all involved organizations and agencies. The proactive, coordinated approach that can be achieved through working together serves to support the mission and goals of both the news media in terms of accurate, comprehensive, professional, and timely coverage, and that of emergency management agencies and emergency service organizations in terms of contributing to the effective, efficient, and safe management of emergency incidents. In addition to the information dissemination and news coverage that occurs throughout the emergency event, through collaboration there is much that these parties can do to prepare the community and its residents in advance of an incident. The various organizations listed as resources in figure 20–5 have available information that can be used by emergency management officials, fire and emergency service organizations, and their news media counterparts, as well as the public, to prepare in advance of the occurrence of a severe weather event or natural disaster. *Are You Ready? A Guide to Citizen Preparedness* is one such resource available through the Federal Emergency Management Agency (FEMA).[1] The various stakeholders of each of these organizations, who at times may find their lives and property in harm's way, expect and deserve nothing less of these organizations that exist to inform, serve, and protect them.

The two job aids in figures 20–4 and 20–5 are designed to serve as resources for fire department and media personnel. Figure 20–4 provides an overview of incident management of weather-related events and natural emergencies, while figure 20–5 provides guidance with respect to media coverage of these incidents.

Incident Management: Weather-Related Events and Natural Emergencies	
Typical Incidents	• Cyclone • Drought • Earthquake • Extreme temperature • Flood • High winds • Hurricane • Ice storm • Lightning • Mudslide • Natural disaster • Severe weather • Snowstorm • Thunderstorm • Tornado • Tsunami • Typhoon • Volcanic eruption • Winter storm
Incident Priorities	1. Life safety 2. Incident stabilization 3. Property conservation
Operational Tactics*	• *Specific information on operational tactics at emergency incidents triggered by severe weather events and/or natural disasters is provided in the earlier incident-specific chapters.* • *See chapter 9 for information on accidents.* • *See chapter 10 for information on building fires.* • *See chapter 11 for information on equipment fires.* • *See chapter 12 for information on hazardous materials fires.* • *See chapter 13 for information on outside fires.* • *See chapter 14 for information on vehicle fires.* • *See chapter 15 for information on hazardous materials incidents.* • *See chapter 16 for information on infrastructure emergencies.* • *See chapter 17 for information on rescues.* • *See chapter 18 for information on searches.*

Fig. 20–4. Incident management job aid. **Note:* The nature of a particular incident will determine the appropriate operational tactics, safety considerations, responding agencies, and resource requirements.

Safety Considerations*	• The priority of life safety of both the public and emergency responders must influence all decisions made and actions taken at these incidents. • The specific safety considerations discussed in each of the preceding incident-specific chapters will be relevant based on the nature of a particular incident resulting from severe weather or a natural disaster. • Personnel should wear appropriate personal protective equipment in accordance with the specific incident, with different PPE being dictated for a building fire that started as a result of a lightning strike, a wildland fire started by that same lightning storm, a water rescue during a flood, or a technical rescue following a structural collapse during a tornado or earthquake. • Personnel and apparatus should always be staged in safe locations. • The dangers associated with rising or moving water must always be considered in the interest of ensuring emergency responder safety. • The dangers that result from the interaction of electricity, gas, and water discussed in chapter 16 must constantly be considered. • Personnel accountability is essential throughout these incidents. • Building collapse, carbon monoxide, confined spaces (limited access, limited ventilation, potentially harmful atmospheres), control zones, critical incident stress debriefing, crowd control, downed power lines, electrocution, evacuation, explosions, hazards, hazardous materials, ignition sources, IDLH atmospheres, perimeter control, personnel accountability, rehabilitation, scene assessment, scene management, search and rescue, shelter in place, stabilization, traffic control, utilities, victims, and weather
Responding Agencies*	• Emergency management agencies, fire department, emergency medical services, law enforcement, hazardous materials teams, technical rescue teams, public works, utility companies, and disaster assistance agencies
Resource Requirements	Personnel* • Emergency management, fire department, emergency medical services, law enforcement, specialized personnel (hazardous materials teams, search and rescue teams), public works, utility companies (electric, gas, sewer, telephone, water), and disaster assistance Personal protective equipment* • Appropriate personal protective equipment should be worn in accordance with the requirements of the specific incident.

Fig. 20–4. Incident management job aid (continued)

	• *Specific guidance on personal protective equipment is provided in the earlier incident-specific chapters.* Apparatus* • *Specific guidance on necessary apparatus is provided in the earlier incident-specific chapters.* Equipment* • Equipment needs will be determined by the nature of the incident. • Traffic control devices will be required at many of these incidents. • *Specific guidance on necessary equipment is provided in the earlier incident-specific chapters.*

Fig. 20–4. Incident management job aid (continued)

Media Coverage: Weather-Related Events and Natural Emergencies	
WHO?	• Number of victims? • Number of and extent of civilian injuries? • Number of and cause of civilian fatalities? • Number of and extent of emergency responder injuries? • Number of and cause of emergency responder fatalities? • Number of trapped persons? • Number of stranded persons? • Number of missing persons? • Personal information on injured individuals? • Personal information on fatalities? • Personal information on trapped, stranded, or missing persons? • *Have appropriate notifications been made?* • *Has information on victims been received from a credible and authorized source?* • *Has information been verified for accuracy?* • How and to which medical facilities were victims transported? • Medical condition of victims (initial and present)? • *See chapter 21 for incidents involving civilian injuries or fatalities.* • *See chapter 22 for incidents involving firefighter injuries or fatalities.*
WHAT?	• Type and nature of severe weather event(s)? • Type and nature of natural disaster? • Similar past severe weather events? • Similar past natural disasters? • Recent weather? • Current weather? • Forecasted weather?

Fig. 20–5. Media coverage job aid

- Direction of weather approach?
- Temperature?
- Humidity?
- Wind direction?
- Wind speed?
- Wind gusts?
- Type of precipitation?
- Amount of rainfall?
- Amount of snowfall?
- Flooding?
- Flood stage?
- Power outages?
- Issues with other utilities?
- Impact on transportation infrastructure?
- Involvement of hazardous materials?
- Damage to buildings?
- Damage to bridges and other structures?
- Damage to vehicles?
- Damage to other property?
- Outdoor storm damage?
- Downed trees and branches?
- Downed electrical power lines?
- Debris-covered roadways?
- Flooded roadways?
- Impassible roadways?
- Responding agencies?
- Roles of agencies?
- Assisting agencies?
- Cooperating agencies?
- Incident priorities?
- Operational tactics?
- Safety considerations?
- Incident management challenges?
- Resource requirements (number and types of personnel, apparatus, and equipment)?
- Resulting health or environmental concerns?
- Life safety dangers associated with the event?
- Recommended actions?
- Warnings or alerts issued?
- Instructions from emergency management agencies?
- Evacuation or shelter in place?

Fig. 20–5. Media coverage job aid (continued)

	• Voluntary (recommended) evacuations? • Mandatory evacuations? • Suggested travel restrictions? • Mandatory travel restrictions? • Road closures or lane restrictions? • Traffic detours?
WHEN?	• Day of week? • Time of day? • Day/night? • Time event began? • Times of specific developments during event? • Time storm made landfall? • Time event arrived in community? • Times specific incidents occurred? • Duration of specific incidents? • Anticipated length of event? • Expected duration of incidents? • Times power outages occurred? • Estimated times power will be restored? • Estimated times other utilities will be restored (gas, sewer, telephone, water)?
WHERE?	• Area affected by event(s)? • Involved communities? • Location of incident(s)? • Nearby recognized landmarks (roads, buildings, etc.)? • Alternate location references (route number vs. street name)? • Proximity to prior events or incidents? • Flooded areas? • Areas of concern for rising waters? • Locations of power or other utility outages? • Electric utility? • Gas utility? • Sewer utility? • Telephone utility? • Water utility?
WHY?	• *See incident-specific chapters for relevant information.*
HOW?	• *See incident-specific chapters for relevant information.*
Sidebar Story Ideas	• Animals in a disaster • Citizens Corps • Climate change

Fig. 20–5. Media coverage job aid (continued)

	• Creating a disaster plan • Cyclones • Dangers of rising water • Disaster assistance • Disaster preparedness • Disaster supply kits • Droughts • Earthquakes • Emergency alerts and warnings • Emergency planning • Extreme temperatures • Evacuation • Evacuation routes • Flood control • Flood insurance • Floods • Hurricane preparedness • Hurricanes • Ice storms • Lightning • Mudslides • Natural disasters • Past natural disasters • Past severe weather events • Recovering from a disaster • Severe weather • Sheltering in place • Snowstorms • Special needs populations • Technical rescue • Thunderstorms • Tornados • Tsunamis • Typhoons • Use of social media during an emergency • Volcanic eruptions • Water rescue • Wildfires • Winter storms
Resources	• American Red Cross (ARC) www.redcross.org • Centers for Disease Control and Prevention (CDC) www.cdc.gov

Fig. 20–5. Media coverage job aid (continued)

- Citizens Corps www.citizencorps.gov
- Federal Emergency Management Agency (FEMA) www.fema.gov
- National Oceanic and Atmospheric Administration (NOAA) www.noaa.gov
- National Weather Service (NWS) www.weather.noaa.gov
- United States Fire Administration (USFA) www.usfa.fema.gov
- U.S. Forestry Service (USFS) www.fs.fed.us
- U.S. Geological Survey (USGS) www.usgs.gov

Fig. 20–5. Media coverage job aid (continued)

Chapter Questions

1. Identify the typical categories of weather-related events and natural emergencies.
2. List the incident management priorities in managing weather-related events and natural emergencies.
3. Discuss several operational tactics related to weather-related events and natural emergencies.
4. Identify the safety concerns related to weather-related events and natural emergencies.
5. Identify the emergency service and support organizations and agencies that are dispatched to weather-related events and natural emergencies.
6. Discuss the resource requirements, in terms of personnel, apparatus, and equipment, associated with weather-related events and natural emergencies.
7. Relate and explain the essential story elements of news coverage of weather-related events and natural emergencies.

Notes

1 Federal Emergency Management Agency. 2002. *Are You Ready? A Guide to Citizen Preparedness*. Emmitsburg, MD: Federal Emergency Management Agency. September.

Chapter 21

Civilian Injuries and Fatalities

Chapter Objectives

- Discuss civilian injury and fatality statistics.
- Identify expectations for media coverage of civilian injuries and fatalities.
- Discuss the importance of professionalism and sensitivity in media coverage of civilian injuries and fatalities.
- Discuss the story elements associated with media coverage of civilian injuries and fatalities.

An essential aspect of the media coverage of any emergency incident is the human impact of the incident. While decisions to evacuate residents or have them shelter in place certainly represent a facet of this human impact, the more significant human impact of an emergency incident is its possible resulting injuries and fatalities. This chapter considers media coverage of civilian injuries and fatalities, while the next examines coverage of firefighter injuries and fatalities. These two chapters are designed to supplement the previous incident-specific chapters and to provide fire department and media personnel with essential guidance on covering such tragedies in a manner that, while meeting the aforementioned stakeholder expectations for media coverage of emergency incidents, also ensures sensitivity in this coverage.

Civilian Fire Injuries and Fatalities

The National Fire Protection Association's (NFPA) annual studies of fire in the United States report that in 2011 there were 1,389,500 fires, resulting in 3,005 civilian deaths and 17,500 civilian injuries. Of those, 2,640 civilian fire deaths and 15,635 civilian fire injuries were in structure fires, 300 civilian fire deaths and 1,190 civilian fire injuries were in vehicle fires, and 65 civilian fire deaths and 675 civilian fire injuries were in outside and other fires.[1] Further analysis revealed that within the United States during 2011, on average one civilian fire injury was reported every 30 minutes, with one civilian fire death being reported every 2 hours and 55 minutes.[2] While contemporary fire departments respond to the numerous other types of incidents considered in the previous chapters, including vehicle accidents that produce significant injuries and deaths each year, the above statistics for civilian fire injuries and deaths illustrate the magnitude of these casualties and thus the importance of the parties to professional and responsible media coverage of these tragedies being prepared to address incident-related injuries and deaths in an appropriate and sensitive manner.

The U.S. Fire Administration in *Civilian Fire Fatalities in Residential Buildings*, a research report published in 2011, examined civilian fatalities in residential buildings between 2007 and 2009. Findings of this study indicated that fires in residential buildings accounted for 81 percent of all civilian fire fatalities and that 91 percent of all civilian fatalities in residential fires involved thermal burns and smoke inhalation. The major human factor contributing to civilian fire deaths was being asleep, with other factors including being physically disabled, possibly impaired by alcohol, and possibly mentally disabled.[3]

Stakeholder Expectations for Coverage of Civilian Injuries and Fatalities

The primary stakeholder groups of the media coverage of emergency incidents and their expectations were introduced in chapter 1 and further considered throughout part 1 of this book. The importance of understanding these stakeholders and their expectations, while essential in the media coverage of any emergency incident, is even more crucial when an incident results in either injuries or fatalities to civilians or fire department personnel. One of the stakeholder groups previously considered was "the public." Earlier, this stakeholder group was addressed in the aggregate. In the interest of fully understanding the responsibilities and challenges of media coverage of emergency incidents involving civilian injuries or fatalities, it is necessary that we identify a subgroup within the public that must be considered in related media coverage. That stakeholder group would be the family and significant others of the victim(s) of an emergency incident, whether an individual was seriously injured or perished through suffering smoke inhalation or thermal burns during a structure fire, or suffered serious injuries or death from a motor vehicle accident (fig. 21–1).

Fig. 21–1. Removal of the victim of an emergency incident

These individuals share the same expectations for media coverage of emergency incidents that were previously discussed. They expect that coverage will be accurate, comprehensive, professional, and timely. While timely information dissemination of the incident details, particularly who was involved, the extent of his or her injuries, and to where he or she has been transported, will be extremely important to a person's family, loved ones, and significant others, hearing about such a tragedy through the news media or second- or third-hand is never appropriate. This is particularly important in the era of social media, in which a victim's loved ones might learn about the occurrence of an emergency incident only after a multitude of others have through extensive interactive social media communications.

Professionalism is obviously a primary expectation in these situations. Waiting until appropriate notifications have been made before releasing the names of victims is an integral aspect of professional coverage of emergency incidents involving civilian injuries or fatalities. The news media frequently report that there have been a number of injuries or fatalities, but do not release names until proper notifications have been made. Accuracy and credibility are likewise important expectations in these situations. As inappropriate and devastating as the premature release of names and other information may be in advance of making proper notifications, irresponsible media coverage that lacks accuracy or credibility can likewise contribute to undue distress on the part of involved individuals. An example of this would be the reporting of a serious automobile accident involving high school students, where media coverage incorrectly reported the death of one of the vehicle's occupants or inaccurate information on who was driving the car or other details of the incident.

Media Coverage of Civilian Injuries and Fatalities

The primary coverage of an emergency incident will relate to the nature of the incident, as examined in the previous incident-specific chapters, and will typically include information on the location, nature, and specifics of the emergency; priorities and goals in managing the incident; operational objectives and tactics; responding agencies; and associated challenges of incident management. However, the occurrence of associated civilian injuries or fatalities will quickly become an aspect of the story that will usually be of interest to the news media and their audience. While responsible reporting is important in all aspects of the media coverage of emergency incidents, it is imperative that release of information about injuries and fatalities be handled in a sensitive and professional manner, with specifics being provided only after appropriate notifications to family and loved ones have been made. Most public safety agencies address this issue by not making specific information available to the media until after these notifications have been properly handled. An example of this would be the release of information to the media and their subsequent dissemination to their readers, listeners, or viewers that "the three occupants of the car were transported to Community Hospital with serious injuries."

While the mission of both parties to responsible media coverage of emergency incidents should always be to ensure the correct reporting of a news story, if there was ever a time that this was absolutely essential it is in the media coverage of emergency incidents involving injuries or deaths. Accuracy, one of the standing expectations of emergency incident media coverage, is thus critical, as is the credibility of the source(s) from and through which information is released. There is perhaps no better example of the use of a PIO as the knowledgeable, authorized spokesperson of a fire or emergency service agency. Only emergency responders who are authorized to speak with the news media should do so, particularly in situations like these. Reporters should likewise understand that they should only seek information from designated and authorized agency representatives. The various public safety representatives at an emergency incident should also respect appropriate boundaries on who should address which issues with the news media, with representatives of law enforcement often handling information dissemination regarding fatalities.

Any information regarding incident victims that is going to be released to and through the news media should be meticulously verified for accuracy prior to its release. Ensuring that prior notifications have been made in advance of the release of specific personal information on victims must be a prerequisite in information dissemination. The information that will most often be incorporated with respect to injuries or fatalities includes the name, town of residence, and destination medical facility in cases where the victim was transported (fig. 21–2).

Professional and responsible media coverage of emergency incidents has been a theme throughout this book and is certainly the hallmark of successful media coverage. While all of the stakeholders of emergency incident media coverage have a right to expect that the information that they receive on breaking news stories such as emergency incidents will be accurate, comprehensive, professional, and timely, the families and loved ones of the victims injured or killed in emergency incidents deserve far more. This unique group of stakeholders deserves to learn about the unfortunate fate of their loved ones in a more personal manner than afforded by radio or television coverage.

Fig. 21–2. Fire department representative briefing the media on incident casualties. (Courtesy of Bill Tompkins.)

While there will always be the tendency to "rush to report" and thus beat news media competitors to bringing the most complete coverage of an incident to their audience, due caution must be exercised when it comes to the reporting of the casualties of an emergency incident. Failure to do so can cause further pain or distress to a victim's family or significant others through the premature release of information, or the release of information that is inaccurate, misleading, or confusing. If there was ever a time to get the story right—the primary responsibility of a news reporter—it is in cases where there are casualties of an emergency incident. This is, once again, a responsibility of all parties that contribute to the media coverage of an emergency incident. Professional, responsible reporting of incident-related injuries and fatalities requires this filtering through the lenses of sensitivity and empathy, in terms of ensuring that the resulting media coverage is handled in the way that you would deserve and expect it to be handled if your loved ones were involved.

The job aid in figure 21–3 serves as a resource for both fire department and media personnel in terms of providing guidance with respect to media coverage of incidents involving civilian injuries and/or fatalities.

	Media Coverage: Civilian Injuries and Fatalities
WHO?	• Number of victims? • Number of individuals who escaped without injury? • Number of injuries? • Nature and extent of injuries? • Number of fatalities? • Cause of fatalities (if known)? • Personal information on injured individuals? • Personal information on fatalities? • Names of victims? • Ages of victims? • Towns where victims resided? • Relationships of victims? • Additional information on victims? • *Have appropriate notifications been made?* • *Has information on victims been received from a credible and authorized source?* • *Has information on victims been verified for accuracy?* • How and to which medical facilities were victims transported? • Medical condition of victims (initial and present)?
WHAT?	• Type of incident? • Responding agencies? • Incident priorities? • Operational tactics? • Safety considerations? • Incident management challenges? • *See incident-specific job aids for further information on incident management and media coverage.*
WHEN?	• Day of week? • Time of day? • Day/night?
WHERE?	• Incident location? • Nearby recognized landmarks (roads, buildings, etc.)? • Past related incidents in same area?
WHY?	• Situational factors contributing to incident? • Contributing human factors? • Were seatbelts worn (accident)? • Presence of smoke detectors (building fire)? • Working condition of smoke detectors (building fire)?

Fig. 21–3. Media coverage job aid

	• Official determination of cause(s)? • Building or fire code violations (building fire)? • Charges brought against responsible parties? • *See chapter 23 job aid for further information on arson and incendiary fires.*
HOW?	• Factors contributing to injuries? • Factors contributing to fatalities?
Sidebar Story Ideas	• Alcohol consumption by vehicle driver/operators • Alternative heater safety • Appliance safety • Careless smoking • Cooking fires • Distracted drivers • Driving safety • Elderly drivers • Fire safety for children • Fire safety for individuals with disabilities • Fire safety for the elderly • Fire safety for the hearing-impaired • Fire safety for the visually-impaired • Holiday fire safety • Home escape plans • Home fire safety • Residential fire fatalities • Residential fire sprinklers • Seatbelt use • Substance abuse by vehicle driver/operators • Smoke alarms • Weather-related accidents • Winter fires • Young drivers
Resources	• Centers for Disease Control and Prevention (CDC) www.cdc.gov • National Fire Protection Association (NFPA) www.nfpa.org • United States Fire Administration (USFA) www.usfa.fema.gov

Fig. 21–3. Media coverage job aid (continued)

Chapter Questions

1. Discuss the civilian injury and fatality statistics presented in this chapter.
2. Relate and explain the stakeholder expectations regarding media coverage of civilian injuries and fatalities.
3. Discuss the importance of ensuring that media coverage of civilian injuries and fatalities is reported in a professional manner.
4. Discuss the important story elements associated with media coverage of civilian injuries and fatalities.

Notes

1 National Fire Protection Association. 2011. *The U.S. Fire Problem*. Quincy, MA: National Fire Protection Association.

2 Ibid.

3 U.S. Fire Administration. 2011. *Civilian Fire Fatalities in Residential Buildings*. Emmitsburg, MD: U.S. Fire Administration.

Chapter 22

Firefighter Injuries and Fatalities

Chapter Objectives

- Discuss firefighter injury and fatality statistics.
- Discuss how a firefighter injury or fatality can involve managing "an incident within an incident."
- Identify expectations for media coverage of firefighter injuries and fatalities.
- Discuss the importance of professionalism and sensitivity in media coverage of firefighter injuries and fatalities.
- Discuss the story elements associated with media coverage of firefighter injuries and fatalities.

An essential element of each and every emergency incident covered by the news media is obviously the dedicated fire department personnel whose responsibility it is to effectively, efficiently, and safely resolve the emergency situation to which they were dispatched. The labor-intensive nature of firefighting and related fire department activities, as well as the inherent hazards and risks of the profession, subject fire department personnel to the potential of injury and death as they respond to, return from, and operate on emergency incident scenes. Although these associated professional risks are acknowledged and serve as a basis for the established priorities in incident management discussed earlier—(1) life safety, (2) incident stabilization, and (3) property conservation—firefighters face the potential of suffering injury or death on a daily basis and unfortunately at times that potential becomes reality. While in recent years about 100 firefighters have died in the line of duty each year, even one such death should be viewed as unacceptable.

Firefighter Injuries and Fatalities

The National Fire Protection Association (NFPA) reported that there were 1,389,500 fires in the United States in 2011.[1] During that year 81 firefighters died while on duty.[2] Longitudinal firefighter fatality research conducted by the U.S. Fire Administration (USFA), based on incident information reported by fire departments using the National Fire Incident Reporting System (NFIRS), has revealed that both volunteer and paid firefighters are the victims of firefighter fatalities, that the fatalities occur while operating on the scene of an emergency incident as well as while en route to or returning from emergency incidents, and that a number of firefighters die each year during training activities. Heart attacks have been found to be the most frequent cause of firefighter line-of-duty deaths. The cause of firefighter fatalities in younger firefighters tends to be traumatic injuries, while medical emergencies represent the primary cause of death in older firefighters.[3]

The U.S. Fire Administration in its 2011 Topical Fire Report Series publication, *Fire-Related Firefighter Injuries Reported to NFIRS*, related that between 2006 and 2008 there were an estimated 81,070 firefighter injuries annually, with 39,715 occurring on an incident scene and 4,880 occurring during response to an incident scene or return to a fire station. The majority of firefighter injuries occur while operating at structure fires. This same study identified the causes of fire-related firefighter injuries in descending order of occurrence as overexertion/strain, exposure to hazard, contact with object, slip/trip, fall, other cause of injury, struck or assaulted, and jump. The nature of fire-related firefighter injuries in descending order of occurrence are strain, wound/bleeding, burns, dizziness/exhaustion/dehydration, asphyxiation/respiratory, fracture/dislocation, cardiovascular, and sickness.[4]

The aforementioned resources of the USFA and the NFPA, as well as additional reports and publications of both organizations, are recommended as valuable resources to fire department and media representatives involved in the media coverage of firefighter injuries and/or fatalities. The National Fallen Firefighters Foundation (NFFF) is likewise a valuable resource to the fire service and to those fire department and media representatives involved in the coverage of a firefighter line-of-duty death (LODD).

An Incident within an Incident—The Rescuer Becomes the Rescued

The overview of incident management provided in chapter 8 emphasized that life safety is a priority that influences every strategic and operational decision on the scene of an emergency incident. The importance of continual size-up throughout the management of an emergency incident was also emphasized in the interest of monitoring both conditions and progress. Together the priority of life safety and the performance of continual size-up contribute to the safety of all firefighting personnel operating on an emergency incident scene. When one considers the number of incidents to which fire departments respond annually and the inherent risks and hazards of firefighting, it is to be expected that a limited number of firefighters will be injured each year and tragically a number will also sacrifice their lives in the line of duty. That having been said, the fire service has adopted the position that it has failed if even one firefighter dies in the line of duty. The National Fallen Firefighters Foundation's *Everyone Goes Home* program is specifically designed to reduce firefighter fatalities through 16 basic initiatives.[5]

When a firefighter experiences an injury on the scene of an emergency incident, such as a structure fire, this occurrence should be immediately communicated to the incident commander to ensure that the firefighter receives immediate assistance, including medical attention if required, and to assign another firefighter to replace him or her in the original assignment. If the injury is minor, such as a firefighter twisting an ankle while carrying a ground ladder from the fire apparatus to the fire building, the above necessary actions to attend to the needs of the injured firefighter and the continuing operational needs of the incident will be relatively transparent, with the firefighter receiving the necessary assistance and care and the assignment of another firefighter allowing incident management and operations to proceed without complication.

Should the injury be more serious, as in the case of the firefighter falling several floors from that same ground ladder, things will obviously be different. The priority of providing needed assistance and medical attention to the injured firefighter will drive the formal and informal actions of firefighting personnel operating on the incident scene. In addition to the personnel assigned by the incident commander to assist the injured firefighter, other fire department members may take it upon themselves to rush to the aid of their colleague. While such actions are both understood and commendable, they can in some circumstances complicate getting the firefighter the medical assistance that he or she needs, as in the case of firefighters crowding a fenced-in area and thus preventing the ready access of emergency medical services personnel or removal of the firefighter to a waiting ambulance.

Depending on the severity of a firefighter's emergency situation and associated injuries, a reasonably good day wherein the fire department is effectively, efficiently, and safely managing an incident can turn into an incredibly bad day almost instantaneously. Interior firefighting crews operating inside a working fire in a commercial building serve to illustrate such a situation. The awareness that one or more firefighters are in actual or possible trouble inside the building could result from not being able to contact the firefighters to establish personnel accountability or from the declaration of a Mayday by the involved firefighter(s). Through issuing a Mayday over the operational radio frequency, a firefighter is able to advise the incident commander and other personnel operating on the incident scene that he or she is in trouble and needs assistance. This may be the result of becoming disoriented or lost within the building, becoming trapped, or running out of breathing air in a self-contained breathing apparatus.

Upon learning that one or more firefighters are in distress, appropriate resources, typically in the form of one or more rapid intervention teams (RITs), will be deployed to locate and assist in getting the lost, injured, or downed firefighter(s) out of the building to a location of safety and necessary medical attention. While much is done on the scene of an emergency incident to ensure the life safety of firefighters, including the appointment of an incident safety officer (ISO), the use of a personnel accountability system, the issuance of personal alert safety system (PASS) devices, and the assignment of rapid intervention teams, there will unfortunately be times that firefighters will perish in the line of duty.

Stakeholder Expectations for Coverage of Firefighter Injuries and Fatalities

The stakeholder expectations for the media coverage of firefighter injuries and fatalities are similar to those for civilian injuries and fatalities discussed in the previous chapter. The expectations that such coverage is accurate, comprehensive, professional, and timely that have been discussed throughout this book certainly apply, as do those related to credibility, sensitivity, and empathy discussed with respect to civilian injuries or fatalities.

The stakeholders of the media coverage of firefighter casualties additionally include the families and loved ones of those who suffered injuries or death, their fire department, and their firefighting colleagues—particularly those on the involved incident scene. The media coverage of such a tragic emergency incident can serve as a fitting tribute to the community's heroes who made the ultimate sacrifice for their community and its residents or call into question their actions and those of their fire department, and thus their individual and departmental reputations.

The fire department will have a vested interest in ensuring that the media coverage meets all of the usual stakeholder expectations, but also that the story is gathered and delivered in a professional manner that demonstrates sensitivity and empathy. The involved firefighters, their families and friends, other fire department members, and the fire department deserve nothing less.

Media Coverage of Firefighter Injuries and Fatalities

While minor injuries will most likely be handled through usual means and communication between reporters and the fire department's PIO, the occurrence of a serious injury or fatality of one or more firefighters will instantaneously invoke certain challenges in terms of managing the original incident, addressing the urgent needs of injured fire department personnel, and ensuring the successful media coverage of both the original incident and the resulting firefighter injuries or fatalities. The incident commander will typically assign an officer to handle the firefighter emergency. If a PIO has not previously been appointed, a firefighter injury or fatality would necessitate such an appointment, given the increased public and media interest that the incident will now generate.

This represents another example of the value of developing a positive working relationship between fire department and media representatives in advance of an emergency incident. The second and third chapters of this book examined the early stages of an emergency incident from the differing perspectives of the fire department and the media. The new incident developments and accompanying challenges resulting from a firefighter emergency will pose a new set of challenges and decisions for the incident commander that he or she will have to address in a timely manner. It will be imperative that reporters on the incident scene have the patience to allow the incident commander the dedicated and undisturbed time required to determine the appropriate actions that must be taken to ensure the life safety of the involved fire department members. It may also take the PIO some time to gather the necessary information to ensure the accuracy of information provided through interviews or press conferences.

The fire department will want to make sure that proper notifications are made before releasing the names of fire department personnel suffering injuries or fatalities. This is important for all of the reasons referenced in the previous chapter on civilian injuries or fatalities. Premature release of information or the reporting of inaccurate, confusing, or misleading information regarding firefighter injuries or fatalities, in addition to causing undue stress for the families and loved ones of involved firefighters and possibly tarnishing the fire department's reputation, can contribute to an enduring hostile working relationship between the fire department and involved reporters and their media organizations. The dissemination of information to the media following a firefighter fatality should be planned and should primarily utilize written press releases and prepared statements delivered in a timely manner at scheduled press conferences.

The news coverage of an emergency incident where a firefighter has been seriously injured or killed does not usually end on the day of the incident. In the event of a firefighter fatality, there will typically be media and community interest in covering the funeral. Useful information on firefighter fatalities, firefighter funerals, line-of-duty death benefits, and support services for surviving families and fire departments can be found on the National Fallen Firefighters Foundation website at www.firehero.org (figs. 22–1 and 22–2).

Fig. 22–1. Firefighter LODD funeral

Fig. 22–2. NFFF memorial service. (Courtesy of National Fallen Firefighters Foundation.)

It has been said that in difficult times the members of a community work together for a common good. Perhaps that is the best way to describe the responsibility that fire department and news media personnel have to ensure that all reporting on firefighter injuries and/or fatalities fully meets the stakeholder expectations of accuracy, comprehensiveness, professionalism, and timeliness. Together, the parties to responsible reporting can ensure that the media coverage of these tragic events is delivered in a sensitive and empathetic manner based on mutual respect that, in addition to ensuring successful media coverage, also brings tribute to the community's public servants and their sacrifice for the community.

The job aid in figure 22–3 serves as a resource for both fire department and media personnel in terms of providing guidance with respect to media coverage of incidents involving firefighter injuries and/or fatalities.

	Media Coverage: Firefighter Injuries and Fatalities
WHO?	• Number of victims? • Number of firefighters who escaped without injury? • Number of injuries? • Nature and extent of injuries? • Number of fatalities? • Cause of fatalities (if known)? • Personal information on injured individuals? • Personal information on fatalities? • Names of victims? • Ages of victims? • Towns where victims resided? • Survivors? • Additional information on victims? • Fire department affiliation of victims? • Company/unit assignment of victims? • Years in fire service? • Years with fire department? • Current fire department rank and tenure in that rank? • Were victims paid or volunteer? • Past fire service positions held? • Other fire and emergency service affiliations? • Funeral arrangements? • *Have appropriate notifications been made?* • *Has information on victims been received from a credible and authorized source?* • *Has information on victims been verified for accuracy?* • How and to which medical facilities were victims transported? • Medical condition of victims (initial and present)?
WHAT?	• Type of incident? • Responding agencies? • Incident priorities? • Operational tactics? • Safety considerations? • Incident management challenges? • Was a Mayday declared? • Were firefighters lost or disoriented within a building? • Were firefighters trapped within a building? • Did firefighters run out of breathing air? • Was there a building collapse (floor, ceiling, or roof)? • Was a rapid intervention team (RIT) utilized?

Fig. 22–3. Media coverage job aid

	• What actions did the RIT take to locate and rescue firefighters? • Were critical incident stress management services provided? • *See incident-specific job aids for further information on incident management and media coverage.*
WHEN?	• Day of week? • Day/night? • Time of dispatch to incident? • Time of arrival on incident scene? • Time of injury? • Time of fatality? • Time of pronouncement?
WHERE?	• Incident location? • Past related incidents in same building or area? • Did injury or fatality occur while responding to an emergency incident? • Did injury or fatality occur while operating on an emergency incident scene? • Did injury or fatality occur while returning to quarters from an emergency incident scene? • Did injury or fatality occur during training? • Did injury or fatality occur while at fire station? • Specific location where firefighter was working at time of accident?
WHY?	• Work assignment of firefighter at time of injury or fatality? • Hazards and risks associated with that assignment? • Was a personnel accountability system utilized? • Was an incident safety officer (ISO) appointed? • Situational factors contributing to injury? • Human factors contributing to injury? • Situational factors contributing to fatality? • Human factors contributing to fatality? • Weather conditions? • Were seatbelts worn (accident)? • Did building construction, condition, or use contribute to injury or fatality? • Presence of building fire protection systems (building fire)? • Working condition of building fire protection systems (building fire)? • Official determination of cause(s)? • Building or fire code violations (building fire)? • Charges brought against responsible parties? • *See chapter 23 job aid for further information on arson and incendiary fires.*

Fig. 22–3. Media coverage job aid (continued)

HOW?	• Factors contributing to injuries? • Factors contributing to fatalities? • Did fatality result from a traumatic injury? • Did fatality result from a medical emergency? • What agencies will be investigating the injury or fatality?
Sidebar Story Ideas	• Building construction • Critical incident stress management (CISM) • Declaring a Mayday • *Everyone Goes Home* initiative • Fire department funerals • Firefighter line-of-duty deaths (LODD) • Fitness for duty • National Fallen Firefighters Foundation (NFFF) • Personnel accountability • Post-incident analysis (PIA) • Post-traumatic stress disorder (PTSD) • Public safety officer benefit program (PSOB) • Rapid intervention teams (RITs) • Tribute stories to local fallen firefighters
Resources	• National Fallen Firefighters Foundation (NFFF) www.firehero.org • National Fire Protection Association (NFPA) www.nfpa.org • National Institute for Occupational Safety and Health (NIOSH) www.niosh.com • Occupational Safety and Health Administration (OSHA) www.osha.gov • United States Fire Administration (USFA) www.usfa.fema.gov

Fig. 22–3. Media coverage job aid (continued)

Chapter Questions

1. Discuss the firefighter injury and fatality statistics presented in this chapter.
2. Explain how the occurrence of a firefighter injury or fatality can create "an incident within an incident."
3. Relate and explain the stakeholder expectations regarding media coverage of firefighter injuries and fatalities.
4. Discuss the importance of ensuring that media coverage of firefighter injuries and fatalities is reported in a professional manner.
5. Discuss the important story elements associated with media coverage of firefighter injuries and fatalities.

Notes

1 National Fire Protection Association. 2012. *National Fire Protection Association Estimates*. Quincy, MA: National Fire Protection Association.

2 U.S. Fire Administration. 2011. *Firefighter Fatalities in the United States*. Emmitsburg, MD: U.S. Fire Administration.

3 Ibid.

4 U.S. Fire Administration. 2011. *Fire-Related Firefighter Injuries Reported to NFIRS*. Emmitsburg, MD: U.S. Fire Administration.

5 National Fallen Firefighters Foundation. 2004. *Everyone Goes Home: Firefighter Life Safety Initiatives*.

Chapter 23
Arson and Incendiary Fires

Chapter Objectives

- Discuss the various roles and responsibilities of fire service personnel in fire investigation.
- Discuss the process utilized in determining the origin and cause of a fire.
- Discuss the process of collecting and preserving evidence.
- Identify the categories utilized to classify the cause of a fire.
- Identify expectations for media coverage of arson and incendiary fires.
- Discuss considerations in the media coverage of arson and incendiary fires.
- Discuss the story elements associated with media coverage of arson and incendiary fires.

The previous incident-specific chapters discussed the operational aspects of managing a variety of fire-related emergency incidents including building fires, equipment fires, hazardous materials fires, outside fires, and vehicle fires. While each of these chapters provided an overview of the actions required to resolve an incident successfully and ensure appropriate media coverage of the incident, they did not consider the possibility that a given fire might be intentionally set rather than occur as a result of a natural or accidental cause. Understanding the process of fire investigation will enable a news reporter to provide more comprehensive coverage of emergency incidents involving fire. The purpose of this chapter is to provide that understanding to members of the news media.

Fire Service Roles in Fire Investigation

In addition to the various roles and responsibilities of the fire department personnel operating on a fire scene discussed earlier, the company officers supervising the firefighting crews are expected to bring to the attention of the incident commander any concern they may have with respect to how a fire may have started. The incident commander in turn will request that a qualified fire investigator be dispatched to the incident scene to determine the origin and cause of a fire, as well as determine if the fire was intentionally started. The fire officers are also responsible for properly securing the fire scene so as to deny access by unauthorized individuals and to protect any evidence that may be present.

Upon arrival on the fire scene, the fire investigator will begin the investigation by talking to the incident commander and interviewing appropriate fire department personnel whose fire suppression activities may have afforded them the opportunity to gain valuable insights regarding the location and behavior of the fire (fig. 23–1). The fire investigator will likewise be interested in learning about the actions taken by fire department personnel that may have impacted the incident scene. The fire investigator will also interview all appropriate individuals, including building residents or occupants, witnesses, and other members of the public that have information that will contribute to the development of an accurate and complete understanding by the fire investigator regarding how the fire started and spread.

Fig. 23–1. Fire investigators being briefed by incident commander. (Courtesy of Bill Tompkins.)

Determining the Origin and Cause of a Fire

Before the fire investigator can accurately determine the cause of a fire, it is necessary to determine its area of origin. The experienced fire investigator will skillfully determine the area of origin by examining such indicators as fire burn patterns, the depth of charring, areas of significant damage, and other physical evidence (fig. 23–2). Evidence that the fire started in multiple locations will obviously be of interest in conducting a fire investigation. Potential causes of a fire include electrical, smoking, incendiary, overheated materials, hot surfaces, open flames, cutting and welding, friction, exposure, spontaneous ignition, combustion sparks, chemical reaction, mechanical sparks, static sparks, lightning, and molten substances. An integral component of a fire investigation is a determination of fire growth and spread from the time of ignition until the time that the fire was first observed by fire department personnel.

Fig. 23–2. Fire investigator examining fire scene. (Courtesy of Bob Sullivan.)

Classifying the Cause of a Fire

The desired outcome of a successful fire investigation is the determination of the cause of the fire. Four classifications are used in identifying the cause of a fire: accidental, incendiary, natural, and undetermined. A fire that started without any deliberate human intervention is considered an *accidental* fire. *Natural* fires result from forces of nature, such as lightning and other weather-related events, with no human intervention contributing to their ignition. Fires that are deliberately started are categorized as *incendiary* fires. When the cause of a fire has not yet been determined, the fire cause is referred to as *undetermined*. It should be noted that experienced fire investigators will refrain from calling fires *suspicious*, as should members of the news media. When an incendiary fire is intentionally and maliciously set in accordance with the criteria articulated in relevant criminal statutes, an act of *arson* has been committed.

Collecting and Preserving Evidence

In cases where it has been determined that a fire was intentionally set and there is the intention of prosecuting the individual who through an act of arson started the fire, the proper collection of evidence will be crucial, as will be the proper securing of the collected evidence, including maintaining the required chain of custody of any and all evidence that is collected. It will be the responsibility of the fire investigator to properly introduce all relevant physical evidence and witness interviews during the legal proceedings involved in the successful prosecution of those who have committed an act of arson.

Media Coverage of Arson and Incendiary Fires

An important component of comprehensive media coverage of a fire incident, regardless of the type of fire, will at times involve reporting on how a fire started, particularly in cases where the fire may not have been caused by an act of nature or an accidental cause. If a house fire was started as a result of a lightning strike during a thunderstorm passing through the area, inclusion of that information in reporting on the incident is certainly desirable. Likewise, a fire investigator may determine that a fire in the same house was started by an improper use of a portable space heater that ignited nearby newspapers and curtains before spreading to the furniture and eventually the house itself.

Only certain fire department personnel will be authorized to release the findings of a fire investigation to the public or the media. Typically, such information will only be disseminated by fire investigators or other members of a fire marshal's office. It will therefore be commonplace for an incident commander or the fire chief to refer any and all questions regarding how a fire may have started to these individuals. There will be times that the nature of an incident may call for involvement of law enforcement personnel, in which case the release of information related to the fire investigation may be handled by authorized law enforcement personnel or a district attorney.

In advance of the occurrence of a fire it is advantageous for reporters to learn who has the responsibility for conducting the fire investigation and the subsequent release of the findings of an investigation within a given jurisdiction. It should be understood that there may be times when it is prudent that fire investigators and/or fire officials classify an incident as "under investigation" or "undetermined" until such time as an accurate determination can be made. The premature, inaccurate, or inappropriate release of information related to a fire or its investigation can in certain situations actually compromise the investigation and potentially undermine the successful prosecution of an individual who intentionally set a fire.

While fire service personnel should always consider all conversations that they have with the news media as being "on the record," it is absolutely essential that any and all information related to an ongoing fire investigation never be inappropriately shared with members of the news media, or for that matter anyone else. At such time as an accurate cause of a fire is determined, that information can be disseminated to the media organizations that provided earlier incident coverage for their use in follow-up reporting.

Additional information on arson, incendiary fires, and fire investigation is available from the U.S. Fire Administration (USFA) at www.usfa.fema.gov and the International Association of Arson Investigators (IAAI) at www.firearson.com. Figure 23–3 provides guidance with respect to media coverage of these incidents.

Media Coverage: Arson and Incendiary Fires	
WHO?	• Owner(s) of involved property? • Operator(s) of involved property? • Occupant(s) of involved property? • Caller(s) reporting incident? • Witness(es) to incident? • Individual(s) questioned? • Individual(s) sought for questioning? • Individual(s) arrested? • Juvenile firesetter? • Number of victims? • *Has information on involved parties been received from a credible and authorized source?* • *Has information been verified for accuracy?* • *See chapter 21 for incidents involving civilian injuries or fatalities.* • *See chapter 22 for incidents involving firefighter injuries or fatalities.*
WHAT?	• Involved property (building, equipment, outside, vehicle)? • Property use? • Occupancy type? • Hazardous materials involved? • Fire detection and alarm systems? • Were smoke detectors operational? • Sprinkler and/or standpipe systems? • Extent of fire damage? • Number of building units involved? • Estimated damage to building? • Estimated damage to contents? • Damage to other buildings or property? • Damage to vehicles? • *See chapters 10 through 14 for incident-specific fire information.*
WHEN?	• Day of week? • Time of day? • Day/night? • Was building occupied? • Were occupants asleep?
WHERE?	• Incident location? • Municipality or neighborhood? • Law enforcement agency? • Past fires in building?

Fig. 23–3. Media coverage job aid

	• Past fires in adjacent or area buildings? • Recent related fires in community? • Past incendiary or arson fires in area?
WHY?	• Factors that contributed to the fire? • Factors that caused the fire to grow and spread? • Contributing human factors? • Weather-related factors? • Problems with building fire protection systems? • Building or fire code violations?
HOW?	• How did the fire start? • Area where the fire started (area of origin)? • Is the fire being investigated? • Agencies involved in fire investigation? • Role of each agency involved in investigation? • Agency handling fire investigation? • Agency handing arrest(s)? • Agency handling prosecution(s)? • Are charges being brought against responsible parties? • Official determination of cause? • Determined to be accidental? • Determined to be incendiary? • Determined to be natural? • Labeled as undetermined? • Still under investigation? • *Do not refer to fire as suspicious.*
Sidebar Story Ideas	• Arson • Arson and the economy • Costs of arson fires • Fire investigation • Incendiary fires • Juvenile firesetting • Motives for arson fires • Prosecution of arson fires
Resources	• Bureau of Alcohol, Tobacco, Firearms, and Explosives (ATF) www.atf.gov • International Association of Arson Investigators (IAAI) www.firearson.com • National Fire Protection Association (NFPA) www.nfpa.org • United States Fire Administration (USFA) www.usfa.fema.gov

Fig. 23–3. Media coverage job aid (continued)

Chapter Questions

1. Relate and explain the roles and responsibilities of fire service personnel in initiating and conducting a fire investigation.
2. Discuss the process of determining the origin and cause of a fire.
3. Discuss the process of evidence collection at a fire scene.
4. Relate and explain the categories used to classify the cause of a fire.
5. Discuss the stakeholder expectations for the media coverage of arson and incendiary fires.
6. Discuss essential considerations in the media coverage of arson and incendiary fires.
7. Discuss the important story elements associated with media coverage of arson and incendiary fires.

Chapter 24

Incidents of National Consequence

Chapter Objectives

- Discuss how local incidents can become incidents of national consequence or interest.
- Discuss the challenges of managing incidents of national consequence or interest.
- Discuss the challenges associated with working with representatives of the national media and their organizations.
- Discuss the strategies that contribute to ensuring successful media coverage of incidents of national consequence or interest.

Although all emergency incidents begin as local events, with an initial response comprised of local emergency responders, the nature, size, and scope of certain incidents will necessitate the response of additional resources from outside the involved community and at times from outside its geographic region. While local and regional resources will routinely be adequate to handle the various types of emergency incidents examined in the earlier incident-specific chapters, there will be times that the resource needs of a particular incident(s) greatly exceed local and regional resources and mutual aid is requested and provided by numerous entities, including state and/or federal agencies and organizations.

Incidents of National Consequence or Interest

Incidents like the terrorist attacks of September 11, 2001, have been referred to as "incidents of national consequence" or "incidents of national significance." By their very nature they also represent incidents of significant interest to the public and thus to the news media responsible for keeping an interested, often anxious, audience updated through accurate, comprehensive, professional, and timely coverage of an incident and its aftermath. While acts of domestic terrorism will always be significant news stories from the standpoint of the news media and the public, there are many other incidents that occur on a more frequent basis that generate significant public, and thus media, interest.

Major weather events or natural disasters such as the massive Northridge earthquake that occurred in Los Angeles in January 1994 or the more recent earthquake that occurred in Virginia and was felt along the East Coast in August 2011 are examples of natural events that generated extensive media coverage. Television viewers witnessed live coverage of the initial impact of the Loma Prieta earthquake that occurred in Los Angeles, California, during the warm-up practice for the third game of the 1989 World Series. Hurricanes and other weather-related events typically produce widespread public and media interest. Extensive news coverage continued for days after Hurricane Andrew, a category 5 hurricane, made landfall in southern Florida and southwest Louisiana in August 1992. Although Hurricane Katrina had decreased from a category 5 hurricane while over water to a category 3 hurricane, when it made landfall in August 2005 and caused severe damage along the Gulf Coast from Texas to central Florida, the extensive news coverage of this event clearly illustrated the ability of media coverage to call attention to the actions of elected and appointed officials, as well as public safety agencies, in the management of an emergency incident. The implications of both positive and negative coverage in terms of reputation management, an important topic discussed earlier, can be significant. The extensive news coverage of Hurricane Irene, which caused extensive flooding and wind damage throughout the East Coast in August 2011, further validates that public and media interest in stories on emergency incidents has perhaps even increased in recent years, as residents and travelers anxiously sought updated information on this dangerous storm, which in many areas produced dangerous winds, extensive flooding, and spawned numerous tornadoes. The extensive media coverage of Hurricane Sandy in 2012 further illustrates the public interest in media coverage of large-scale weather events.

While the great majority of emergency incidents to which fire departments respond will be fairly routine and only of local public and media interest, some incidents will instantaneously become newsworthy to national media organizations. While every day fire departments respond to structure fires with no fanfare or media interest beyond local media organizations, a fire in a known landmark, at a celebrity's home, or a tragic fire resulting in multiple deaths will quickly change the level of public and media interest. The tragic fire that occurred in the Station Nightclub in West Warwick, Rhode Island, in February 2003 and resulted in 100 civilian fatalities is an example of one such event. While wildland fires are frequent occurrences in certain areas, the fire that occurred in Oakland Hills, California, in October 1991 that resulted in 25 deaths and the destruction of 2,900 structures is another example of an emergency incident of great interest to the public and the news media.

Although transportation accidents routinely occur throughout the nation, resulting in fire, rescue, or hazardous materials incidents, certain incidents based on their location, magnitude, or impact instantaneously become newsworthy stories to the national media. The derailment and subsequent fire of a freight train carrying hazardous materials in the Howard Street Tunnel in Baltimore, Maryland, in July 2001 was one such incident. This incident, which lasted for a number of days, disrupted freight trains throughout the Northeast corridor, as well as forced the cancellation of three scheduled games of the Baltimore Orioles. Interestingly, one national television network incorporated brief coverage of a hazardous materials incident involving a fire in a tank truck that occurred the same day in Chester County, Pennsylvania. This incident, which was effectively, efficiently, and safely managed by local emergency responders, was covered by a local network affiliate and later picked up for inclusion in the network's evening news broadcast. The airplane crash near Buffalo, New York, in February 2009 that killed 50 people is another example of a breaking news story that drew national attention. Although the many accidents that occur on Interstate 78 in Pennsylvania, particularly during inclement weather, typically receive only local media coverage, the fact that a 50-mile stretch of this primary roadway was shut down for an extended period of time by the Valentine's Day blizzard of 2007 and received extensive national media coverage likewise illustrates how a local or regional event can quickly assume a posture of national interest.

Managing Incidents of National Consequence or Interest

While the incident management process and the National Incident Management System (NIMS) will be utilized in the management of any and all incidents, incidents of national consequence and/or those that capture national public or media interest present additional challenges to those responsible for managing an incident in an effective, efficient, and safe manner. While the scope of this book does not provide for an extensive discussion of the many command and operational issues of managing an incident of national consequence, fire service and media personnel can further their understanding of these challenges through reading *Responding to Incidents of National Consequence*, a publication prepared by the U.S. Fire Administration after September 11, 2001.[1]

A key finding of this research was that the size, scope, or complexity of an incident may overwhelm local emergency resources. A theme that threads throughout the issues identified with respect to the preparation before an incident, the initial response to an incident, the ongoing incident stabilization activities, and the post-event and recovery activities is the importance of providing information to the public. The report states, "The local jurisdiction will be inundated with requests from the public for information," highlighting both the importance and urgency of providing timely updates to the public and the media. Incidents of national consequence and/or public interest thus represent a significant challenge in terms of responding to stakeholder expectations regarding the dissemination of accurate, comprehensive, professional, and timely information. This is obviously a prime example of a situation where the local news media can partner with fire and emergency service organizations as well as government officials to ensure that this informational need is properly addressed. The positive working relationships and partnerships discussed in part 1 of this book serve as the foundation for successful media coverage and the dissemination of vital and often time-sensitive information to the public.

The National Players Take the Stage

Just as the initial response to an emergency incident will consist of local resources from the fire department and other emergency service organizations, so too will the initial media representatives on an emergency incident scene be the local reporters with whom fire department personnel interact on regular basis. In managing the incident the incident commander will request additional mutual aid resources and that appropriate notifications be made to state and/or federal agencies having jurisdiction, such as the National Transportation Safety Board (NTSB) in the case of an aircraft accident or a train derailment or the Environmental Protection Agency (EPA) in the event of a significant hazardous materials release.

Local news media representatives will likewise call their newsrooms to request that additional resources such as reporters, photojournalists, news vans, or news helicopters be dispatched to assist in coverage of the incident. If their media organizations are affiliated, they will alert the national news media of the story that they are covering in the interest of making their incident coverage available to them. The local television station making available their ground and aerial coverage of the earlier mentioned tank truck fire for use by their network would be an example of such information and coverage sharing. Upon learning of an incident, regional and national media organizations outside the local coverage area make the determination of how newsworthy a story is to their audience and thus determine whether to deploy resources to cover the story. While the national coverage may simply incorporate local coverage or a network television anchor talking live to a local reporter on the incident scene, it is customary to actually send resources to the scene of a major emergency incident.

Ensuring Successful Media Coverage

As well as being inundated by requests from the public during one of these incidents, it is to be expected that the news media will be even more persistent, and at times even demanding, in getting the needed information to cover the story of an emergency incident successfully. Ideally, their intentions with respect to getting the story right and reporting in a manner that fully meets stakeholder expectations will correspond with those of the local fire department and media professionals discussed throughout this book. It should be noted that the perspective of the national media will often differ from that of the local news media in terms of the nature of their coverage. Whereas a local television station will incorporate a great deal of updates, advisory information, and warnings for the benefit of a community and its residents, the national media will more likely focus on the nature, size, scope, and aftermath of an emergency. They will also be interested in incorporating visual images that depict the magnitude of the tragedy, such as images of a town that is underwater, and human interest stories, such as interviews with residents who lost their homes or fire and emergency service personnel involved in making rescues.

Just as the resource needs of an incident may overwhelm the available local resources in a community, necessitating requests for mutual aid resources, the arrival of the national news media to the scene of an emergency incident of national interest can likewise present many logistical challenges that must be anticipated and planned for in the early stages of incident management. The downing of the fourth plane in Shanksville, Pennsylvania, during the September 11 terrorist attacks triggered a massive deployment of regional and national news reporters and photojournalists who brought the tools and equipment of their profession, including news vans and satellite trucks, to this remote area that most of them had previously not even heard of. The events of that day clearly put this small community on the map for an army of reporters and news media professionals.

While the primary role and responsibilities of an incident commander at an incident of national public and media interest will obviously be to focus on the effective, efficient, and safe resolution of the emergency situation, it is imperative to remember that ensuring that necessary and appropriate information is provided to the public and the media is also an essential responsibility within the command function. Although major, high-visibility incidents like these have the potential of presenting unprecedented challenges to those responsible for managing the incident, such as the incident commander in Shanksville, the principles of successful incident management that derive from sound management theory and practice and are incorporated into the NIMS have demonstrated their value in contributing to successful incident management, including the provision of necessary and appropriate information to the public and the media.

These incidents obviously necessitate the appointment and empowerment of a highly qualified PIO and in some cases the assignment of additional personnel to assist with public information and media relations. In addition to designating an appropriate media area from which reporters can conduct interviews and report on the story while photojournalists capture vivid images that further tell the story, a staging area for the massive convoy of media vehicles that may make the pilgrimage to the incident scene should be established and properly managed. The establishment and management of control zones that prevent unauthorized persons, including media representatives, from wandering around the incident scene must be a priority in light of the fact that the agenda of every reporter and photojournalist on the incident scene will be to get something that competing media organizations do not have and to report on aspects of the incident before the competition does.

While the challenges of working with an unknown cast of news media characters, including well-known reporters who on occasion will have demanding personalities, can likewise be overwhelming, having the right fire department personnel assigned to handle media relations will contribute to a successful outcome in terms of cooperative working relations and successful media coverage of the emergency incident. Local media personnel who have good working relationships with both the fire department and their visiting media colleagues can often serve in boundary spanning roles that reduce the potential for ambiguity and conflict.

Chapter Questions

1. Relate and explain how emergency incidents that occur at the local level can become incidents of national consequence or interest.
2. Discuss the challenges of managing an incident of national consequence or interest.
3. Discuss the challenges that may arise in working with national media organizations and their personnel.
4. Discuss the strategies that contribute to successful media coverage of incidents of national consequence or interest.

Notes

1 U.S. Fire Administration. 2004. *Responding to Incidents of National Consequence.* Emmitsburg, MD: U.S. Fire Administration.

Chapter 25
Final Thoughts

Chapter Objectives

- Discuss the importance of collaboration, cooperation, and professionalism in emergency incident media coverage.
- Discuss the roles of emergency service and media professionals in fulfilling their shared responsibility for accurate, comprehensive, professional, and timely media coverage of emergency incidents.

It would seem that this final chapter represents the end of a journey that you have traveled as a member of the fire service or the news media. While in a sense that is true, in reality it represents only a milepost in your pilgrimage toward enhancing the media coverage of emergency incidents that fall within the scope of your professional responsibilities. Whether you represent a fire department or a news media organization, it is your knowledge, skills, and attitudes with respect to media coverage of emergency incidents that will determine the success of the resulting coverage. This book, written for both fire service and media professionals, was developed to serve as a tool designed to prepare both parties to the media coverage of emergency incidents—the PIO and reporters—to more successfully enact their roles and responsibilities with respect to the news coverage of these incidents. Similarly, it was also intended to serve as a useful reference during the media coverage of a particular emergency incident.

Successful media coverage of an emergency incident requires that both parties to that coverage fully understand their respective roles, responsibilities, perspectives, and challenges with respect to the effective, efficient, and safe resolution of the emergency situation, and the provision of associated media coverage that meets the expectations of stakeholders that coverage be accurate, comprehensive, professional, and timely. Successful media coverage of emergency incidents is in reality a team sport, in many ways similar to the coordinated actions required of fire and emergency service personnel to resolve an emergency situation. Through working together, the primary representatives of the fire department and the news media—the PIO and reporters—can develop the positive working relationship and implement the synergistic actions necessary to advance the media coverage of emergency incidents to the exemplary level that not only fully meets, but exceeds, the expectations of all of the stakeholders of this news coverage, including the organizations that they represent.

On a number of occasions throughout the book, the word "professional" was incorporated into such phrases as "fire department and media professionals." That word was specifically chosen to emphasize the vital role that the PIO and news reporters play in contributing to the delivery of the outstanding coverage of emergency incidents that stakeholders both expect and deserve. Just as the professionalism and collaboration of the incident commander and the incident safety officer contribute to the overall safety on an incident scene, so too does the collaboration, cooperation, and professionalism of fire department and media representatives determine the success of the media coverage of the incident.

The importance of assigning public information and media relations responsibilities to a designated PIO possessing the appropriate qualifications and interest in serving in this important role bears repeating, as does the merit of establishing a positive working relationship between the parties to media coverage. When both parties are enthusiastic about the roles that they play and enact these roles with dedication and professionalism, enhanced coverage will be within their reach (fig. 25–1).

The challenge that remains as you complete your reading of this book is to determine the appropriate next steps that you and your counterparts to successful media coverage must take if you are truly committed to achieving excellence in the media coverage of the emergency incidents that occur within your community. The strategies discussed in part 1 of this book were intended to help you chart the future course of your continuing pilgrimage or quest to ensure that the media coverage of every emergency incident fully meets and ideally exceeds the expectations of the various stakeholders discussed earlier. You and those you partner with to ensure the accuracy, comprehensiveness, professionalism, and timeliness of every story relating to an emergency incident delivered to newspaper readers, radio listeners, television viewers, and a growing audience who get their news from the Internet and through social media comprise one final stakeholder group that has both a vested interest and high expectations for the resulting media coverage.

I trust that the incident-specific information provided in part 2 proved useful in enhancing your knowledge and understanding of the elements and strategies that contribute to the successful resolution of emergency situations by fire and emergency service personnel, as well as serving as job aids that will serve you well as you enact your roles and responsibilities on the scene of an emergency incident. Your individual success, as well as that of the fire department or media organization that you represent, all comes down to your preparation, the development of the necessary collaborative working relationships, and the dedication and professionalism with which you undertake your work.

Fig. 25–1. Fire commissioner interacting with the news media. (Courtesy of Philadelphia Fire Department.)

Throughout the book reference was made to preparing fire department and media professionals, and I trust that you will agree that the book has and will continue to contribute to your professional preparation and success. When you think about it, while the intended audience of the book is those individuals who currently serve as or aspire to serve as a PIO or news reporter, and both audiences have been properly portrayed as professionals within their own industry, the reality is that achieving the enhanced level and quality of media coverage advocated throughout this book will be premised on both parties to the media coverage of emergency incidents, regardless of whether they work for a fire department or a news media organization, becoming "media professionals."

Chapter Questions

1. Discuss the importance of collaboration, cooperation, and professionalism in the media coverage of emergency incidents.
2. Discuss the roles and responsibilities of the parties to successful media coverage.
3. Discuss how emergency service and media professionals can ensure that media coverage of emergency incidents fully meets, and ideally exceeds, stakeholder expectations.

APPENDICES

Eight appendices are provided as resources for use by media personnel in their reporting on emergency incidents. Appendices A and B are provided in the interest of relating the terminology and acronyms used by fire and emergency service personnel during emergency incident operations. These appendices serve to enable reporters and other media personnel to enhance their understanding of emergency incident operations through clarification of the technical language and jargon used by members of the fire service, thus addressing an issue known to compromise communication effectiveness and understanding.

The successful resolution of an emergency incident requires the development and implementation of appropriate tactics on the emergency incident scene. The resources that are utilized in incident management fall into three categories: personnel, apparatus, and equipment. An overview of the specific resources that fall under each category is provided in appendices C, D, and E.

While the initial response to all emergency incidents will involve local fire and emergency service organizations, major incidents, in addition to requiring mutual aid assistance from other fire and emergency service organizations, will often trigger the response of state and federal agencies and resources. Appendix F provides an overview of federal agencies that may have a role to play in a given emergency incident, based on its nature and scope. This appendix also delineates the roles of several relevant private organizations.

The pivotal role of the public information officer (PIO) was discussed throughout this book. While the roles and responsibilities of a PIO were discussed in chapter 4, appendix G provides a sample position description for this essential position in the successful media coverage of emergency incidents.

The responsibilities of the incident commander and PIO within the context of incident management were examined in chapter 8. Appendix H provides an overview of the National Incident Management System (NIMS) and delineates the responsibilities of the command and general staff positions not considered earlier. The responsibilities of the command staff positions of safety officer and liaison officer, as well as the general staff positions of operations, planning, logistics, and finance/administration are delineated.

Appendix A

Emergency Service Terminology

This appendix is provided to enhance the understanding that members of the news media have of the terminology used by fire and emergency service personnel during emergency incident management and operations. While this listing of terminology is fairly extensive, it does not include terms relating to emergency service positions, apparatus, and equipment; those terms are listed in appendices C, D and E, respectively.

absorption. Passage of a substance into the body through the skin.

action plan. Plan developed to manage an emergency incident (also called an incident action plan).

advanced life support (ALS). Advanced level of medical treatment, including airway management and administration of medication.

air attack. Use of aircraft in the application of fire extinguishing agents during wildland firefighting.

aircraft accident. Aircraft incident resulting in injury, death, or damage.

aircraft incident. Situation with the potential of compromising the safe operation of an aircraft.

aircraft rescue and fire fighting (ARFF). Operations related to managing aircraft accidents and incidents.

arson. Criminal act of intentionally and maliciously starting a fire.

arsonist. Individual who commits an act of arson.

assisting agency. Agency providing personnel and other resources to the agency responsible for managing an incident.

authority having jurisdiction (AHJ). The legal entity with statutory authority to enforce codes.

automatic aid. Prearranged aid from other agencies and organizations that is dispatched automatically based on dispatch protocols.

automatic alarm. Dispatch resulting from an activation of building detection systems and transmittal of an alarm.

automatic sprinkler system. Extinguishing system within a building that activates and discharges water in the event of a fire.

backdraft. Explosion of superheated gases occurring when oxygen is introduced into an oxygen-deprived area.

basic life support (BLS). Basic level of medical treatment that involves airway, breathing, and circulation management.

battalion. A number of fire department stations within a geographic area.

blood-borne pathogen. Microorganisms present in human blood and other body fluids that can result in the transmission of disease.

boiling liquid expanding vapor explosion (BLEVE). Rapid vaporization of a liquid following the catastrophic failure of its container.

bomb squad. Agency with specialized training and equipment for the handling of explosive devices.

breathing air. Atmospheric breathing air that is filtered, compressed, and stored for use in self-contained breathing apparatus and other respirator devices.

building code. Regulations regarding the construction, maintenance, and renovation of buildings adopted by law.

building survey. Survey of a building conducted by fire department personnel during preplanning activities.

cache. Complement of tools, equipment, or supplies available for use during an incident.

canine search. Utilization of specialized trained search dogs in locating incident victims.

carbon monoxide (CO). Deadly gas produced by incomplete combustion that is odorless and colorless.

cause. Initiating factor resulting in the occurrence of a fire.

certification. Process utilized to validate the knowledge and skills of fire and emergency service personnel.

Class A fire. Fire involving ordinary combustibles.

Class B fire. Fire involving flammable or combustible liquids or gases.

Class C fire. Fire involving energized electrical equipment.

Class D fire. Fire involving combustible metals.

Class K fire. Fire involving combustible oils in cooking appliances.

command. Designation used to identify the incident commander at an emergency incident.

command post (CP). Designated location from which the incident commander and command staff enact their responsibilities during an incident.

communications center. Specialized facility equipped to receive requests for emergency assistance from the public, dispatch necessary resources, and support emergency communications (also called a dispatch center).

company. Fire department organizational unit composed of a fire apparatus and assigned personnel.

conduction. Transfer of heat as a result of direct contact.

confined space. Area having limited access, limited ventilation, and a potentially harmful atmosphere.

confinement. Preventing the spread of fire from an involved to an uninvolved area or controlling the flow of a spill.

conflagration. Large fire involving many buildings that transcends natural boundaries, such as city streets or rural roads.

containment. Stopping the release of a material from a container.

control zone. Zone established on an incident scene to restrict access and limit exposure to hazards. Three zones are typically established: a hot zone (restricted), a warm zone (limited access), and a cold zone (support).

convection. Transfer of heat through heated liquids or gases.

cooperating agency. Agency or organization supplying assistance other than direct operational or support functions or resources in the management of an emergency incident.

critical incident stress debriefing (CISD). Counseling provided to emergency responders to enable them to deal with the emotional and psychological trauma of an incident.

crowd control. Actions designed to limit public access to an emergency incident scene.

declared emergency. Aircraft emergency where the crew has notified an airport that they have an emergency situation.

decontamination. Removal of hazardous materials from personnel, equipment, and apparatus.

defensive attack. Firefighting strategy or mode wherein firefighting efforts take place from the outside of a building in an attempt to control the spread of fire to adjacent exposure buildings (also called an exterior attack).

dike. Action taken to contain liquids.

dispatch. Assignment of fire department and other resources to respond to an emergency incident.

district. Geographic area to which a particular fire department responds on an initial alarm (also called a response district).

draft. Acquisition of water from a static source for use in firefighting operations.

elevation drawing. Drawing that shows the elevations and grading of the various sides of a building.

emergency operations. Activities involved in responding to and managing an emergency incident.

emergency operations center (EOC). Facility established by a governmental entity to support the management of an emergency incident or incidents.

emergency operations plan (EOP). Plan developed and maintained by a jurisdiction that addresses the management of hazards.

Emergency Response Guidebook (ERG). Reference manual provided by the U.S. Department of Transportation to assist emergency responders in the identification of hazardous materials and determining appropriate initial actions that should be taken at a hazardous materials incident.

engine company. Firefighters and their assigned fire pumper responsible for water supply and fire suppression activities.

Enhanced 9-1-1. 9-1-1 system that provides for the automatic identification of callers and their location.

evacuation. Removing exposed individuals from a hazardous location.

explosion. Chemical or physical process resulting in rapid release of gas into an environment.

explosive atmosphere. Atmosphere that falls within the explosive limits for a specific material.

exposure. Adjacent areas or surrounding buildings to which a fire could potentially spread.

exterior exposure. Building or object to which a fire could extend.

extinguish. To completely put out a fire.

extremely hazardous substance. Chemicals that the Environmental Protection Agency has identified as having the potential of being extremely hazardous in the event of a release or spill.

extrication. Process of removing a victim trapped by equipment or machinery.

fill site. Site where tankers or tenders are loaded with water during water shuttle operations.

fire alarm system. System designed to detect the presence of a fire, alert building occupants, and report the fire.

fire brigade. Personnel within an industrial facility trained and equipped to handle fire suppression activities.

fire cause determination. Process of determining the origin and cause of a fire.

fire department connection. Connection through which a fire department can connect to a building's sprinkler and/or standpipe system to increase the volume and pressure of water delivered by the system.

fire detection system. System designed to detect and report the occurrence of fire.

fire hazard. Act, material, or condition that contributes to the start or growth of a fire.

fire hydrant. Appliance connected to a system of water supply mains to which a fire department can connect as a water supply source.

fire line tape. Plastic tape used to establish and enforce hazard control zones at an emergency incident.

fire prevention. Activities designed to prevent fire through educating the public on hazards.

fire pump. Pump designed to supply water to building fire protection systems.

fire stop. Building features designed to limit the horizontal and vertical spread of fire.

fire suppression. Actions that relate to and support the extinguishment of a fire.

first alarm. The resources and units initially dispatched to an emergency incident.

flammable. Property wherein a material is capable of burning.

flashover. Situation where all objects in a given room have reached their ignition temperature and fire involving all objects instantaneously occurs.

floor plan. Drawing showing the layout and use of each floor within a building.

foam blanket. Layer of foam applied to a surface that is on fire to extinguish it through denying the oxygen necessary to sustain the fire.

forcible entry. Actions of emergency response personnel designed to gain access into secured areas.

fully involved. Term used to describe a situation where an entire building is involved in fire.

general alarm. Large emergency incident to which all or the majority of a jurisdiction's resources are committed.

hazard. Condition or substance that presents a danger or source of risk.

hazard area. Area on an incident scene to which entry is denied to all but essential personnel.

hazard class. Designation of hazardous materials into categories based on their properties.

hazardous material. Material that presents a risk to health, safety, and the environment in those situations where it is not properly controlled.

hazardous material incident. An emergency incident involving the potential or actual release of a hazardous material.

heat detector. Detection device designed to monitor increases in temperature.

heat exhaustion. Illness caused by exposure to excessive heat.

heat stroke. Illness caused by exposure to excessive heat that compromises the body's normal heat regulating mechanisms.

high-angle rescue. A rescue conducted at a higher elevation necessitating securing of rescuers with ropes to ensure their safety.

horizontal ventilation. Process through which heat and smoke are horizontally removed from a building through doors and windows.

hose lay. Hose connecting a water supply source to a fire pumper on the incident scene.

HVAC system. Mechanical system within a building that provides for heating, ventilation, and air conditioning.

hydraulic ventilation. Use of a fog stream created by a hoseline in the movement of air and smoke from a window of a building.

ignition source. Mechanism or source that initiates combustion resulting in a fire.

immediately dangerous to life and health (IDLH). Atmosphere that represents an immediate life hazard or is capable of producing irreversible health effects.

implosion. Rapid inward collapse of a building.

incendiary device. Device or mechanism designed to start a fire.

incendiary fire. A deliberately set fire.

incident. A situation requiring the services of a fire or emergency service organization(s).

incident action plan (IAP). Plan developed for use in the management of an emergency incident.

incident command post (CP or ICP). Location from which the incident commander and command staff manage an incident.

incident command system (ICS). System designed and utilized to contribute to the effective, efficient, and safe management of an incident. Also called an incident management system (IMS).

incident management team (IMT). Incident commander and appropriate command and general staff assigned to an incident.

industrial fire brigade. Personnel within an industrial facility trained and equipped to handle fire suppression activities.

ingestion. Passage of a substance into the body through the mouth.

inhalation. Passage of a substance into the body through the nose or mouth.

initial alarm. Resources initially dispatched to an emergency incident.

initial attack. Actions taken by the first arriving units on a fire scene.

interior exposure. Uninvolved areas of a fire building that are susceptible to the spread of fire.

investigation. Process through which the cause of a fire or other emergency is determined.

joint information center (JIC). Facility established to coordinate all dissemination of incident information.

joint information system (JIS). System designed to ensure that the release of incident information is consistent, coordinated, and timely.

ladder company. Firefighters and their assigned ladder truck responsible for forcible entry, laddering a building, ventilation, search and rescue, utility control, salvage, and overhaul.

landing zone. Area designated for the landing of a helicopter on an emergency incident scene.

lightweight construction. Construction incorporating lightweight building materials, such as trusses, that are subject to early loss of integrity and failure when subjected to fire.

line-of-duty death (LODD). Emergency responder fatality resulting during performance of official duties.

mass casualty incident (MCI). Emergency incident involving multiple injuries or fatalities.

master stream. Water supply appliance capable of delivering a large volume of water in firefighting operations.

material safety data sheet (MSDS). Documentation provided by a chemical manufacturer that delineates relevant information about a chemical, its properties, and appropriate actions in the event of a fire, release, or spill.

Mayday. Recognized distress signal utilized to declare that a firefighter is in danger.

mechanical systems. Equipment systems present in a building.

mechanical ventilation. Means of ventilation that involve the use of devices such as fans and smoke ejectors.

mitigation. Actions taken to reduce or eliminate the actual or potential harmful effects or consequences of an incident.

mobile attack. The use of fire apparatus using "pump and roll" capabilities during wildland firefighting.

motor vehicle accident (MVA). Emergency incident wherein a motor vehicle strikes another vehicle or a stationary object.

multiagency incident. Incident in which additional agencies assist.

mutual aid. Additional resources requested by an incident commander that respond to assist at an emergency incident under a previously agreed to reciprocal arrangement.

National Fire Incident Reporting System (NFIRS). Incident reporting system utilized throughout the United States that provides for the documentation of relevant information regarding incidents.

National Incident Management System (NIMS). Incident management system, developed by the U.S. Department of Homeland Security, designed to contribute to the effective, efficient, and safe management of emergency incidents.

natural ventilation. Technique utilizing the wind to ventilate a building through windows and doors without the use of any mechanical devices.

negative pressure ventilation. Ventilation technique using smoke ejectors and other mechanical devices to pull smoke from a building.

nonintervention strategy. Allowing a fire involving certain hazardous materials to burn rather than extinguish it based on a determination that this is the prudent action.

occupancy. The current use of a particular building.

occupancy classification. A classification system based on building use that is incorporated into building and fire codes.

offensive strategy. Firefighting strategy or mode wherein firefighters engage in an aggressive interior attack intended to locate, confine, and extinguish a fire (also called an interior attack).

operational period. Period of time during which a given set of operational actions specified in an incident action plan will be executed.

operational strategy. Overall plan for the resolution of an emergency situation.

operational tactic. Specific actions required in support of the strategic goals for the incident.

origin. The area or location where a fire originated.

outside aid. Assistance received from organizations and agencies outside a jurisdiction when no mutual aid arrangements exist.

overhaul. Actions undertaken after fire extinguishment to check for remaining fire in hidden areas.

pancake collapse. A building collapse where collapsing upper floors drop onto lower floors, often causing a cascading effect.

perimeter control. Establishing and enforcing boundaries around an incident in the interest of restricting access.

personnel accountability report (PAR). Conducting a roll call to account for all personnel during an emergency incident.

placard. Marking system used on buildings and in transportation to enable emergency responders to identify the primary class of an involved hazardous material(s).

planning meeting. Meeting conducted at various points during the management of an incident to review progress and plan for an upcoming period.

plot plan. Drawing showing the layout of buildings on a plot of land.

positive pressure ventilation (PPV). Ventilation technique wherein a sufficient volume of air is blown into a confined space so as to displace air through a building opening.

post-incident analysis (PIA). A review and analysis of an incident by members of all responding organizations in the interest of gaining insights regarding incident management and operations.

pre-incident plan. Written document that details building information including construction, occupancy, protection systems, and exposures, as well as planned operational strategies for handling emergencies (also called a preplan).

preplanning. Process of advance preparation whereby a fire department develops an understanding of buildings within its response area, as well as appropriate strategies for managing incidents at these facilities.

product identification number. Number utilized in the Emergency Response Guidebook to identify specific product names.

public safety answering point (PSAP). Facility where requests for emergency assistance are received.

radiation. Energy emitted from a radioactive source in the form of particles or waves.

rapid intervention team (RIT). Team of firefighters that are equipped and immediately deployable to rescue interior firefighters in the event of an emergency situation, such as a firefighter becoming trapped, lost or disoriented, or running out of breathing air. Also called a rapid intervention crew (RIC).

recovery. Situation where a victim is dead and the focus of efforts changes from attempting to locate and rescue the victim to recovery of the body.

rehabilitation. Activities performed on an incident scene to allow response personnel to rest and rehydrate as appropriate (also referred to as rehab).

rekindle. Reignition of a fire after it has been extinguished.

relay operation. Placing additional pumpers in a hose lay in situations where the supply line covers an extended distance and additional pumpers are required to ensure the delivery of an adequate water supply.

relocate. Temporary deployment of fire apparatus and personnel to other stations to provide for the coverage of additional calls during a major incident that has depleted resources (also called a standby).

rescue. Removing a victim to a safe environment.

rescue company. Firefighters and their assigned rescue apparatus responsible for rescue and extrication activities.

resource. Personnel, apparatus, equipment, facilities, and supplies available for use during the management of an emergency incident.

resource management. Process whereby available resources are utilized in an effective, efficient, and safe manner.

response time. Interval between the time of dispatch and arrival on an incident scene.

restricted zone. Established control zone on an incident scene into which only essential personnel involved in a particular operation are permitted.

rollover. Situation where unburned superheated gases in a confined area reach their flammable range and with introduction of oxygen ignite, resulting in the fire rolling across the ceiling of the room.

salvage. Activities performed during firefighting operations designed to reduce further property damage resulting from water or smoke.

scene assessment. Evaluation of the hazards associated with operations on an incident scene.

scene management. Incident management activities that involve ensuring the life safety of personnel operating on an incident scene.

search. Activities conducted to determine the location of victims.

search and rescue. Operations related to locating and rescuing victims.

shelter in place. Strategy of having occupants remain in safe and protected areas of a building, rather than evacuating.

shipping papers. Documents that provide information on specific products and cargo in transportation shipments.

single command. Command structure where one individual serves as the incident commander.

size-up. Mental evaluation performed at the start of an incident and at various points during the incident that enables the incident commander to make informed decisions regarding the management of the incident.

smoke. Visible products of combustion produced by a fire.

smoke control system. Building system designed to control the movement of smoke within a building.

smoke detector. Device that detects the presence of products of combustion and alerts building occupants.

spontaneous combustion. The self-ignition of a material as a result of an internal chemical or biological reaction (also called spontaneous ignition).

sprinkler. Device that is designed to activate when a specified temperature is reached and discharge water in its immediate area.

sprinkler system. System incorporating a water supply, piping, and sprinkler heads that is engineered to address the fire protection needs of a building.

stabilization. Process of supporting objects so they will remain in their existing position, thus facilitating the extrication and rescue efforts of emergency responders.

stack effect. Dynamic created when, upon the occurrence of a fire, a building acts as a chimney for heat and smoke from lower floors to escape to upper floors.

staging. Process whereby responding units report to a designated area and await specific assignments and instructions.

staging area. Area on an incident scene designated for the staging of arriving apparatus.

standard operating procedure (SOP). Procedure that delineates how specific operations, such as establishing a water supply, conducting a search, or ensuring personnel accountability on an incident scene, are to be conducted. Also called a standard operating guideline (SOG).

standby. Temporary deployment of fire apparatus and personnel to other stations to provide for the coverage of additional calls during a major incident that has depleted resources (also called a relocate).

standpipe hose. Hose that is preeconnected to a standpipe system.

standpipe system. System of piping, hose connections, and standpipe hoses installed in a building to support firefighting operations.

strategic goal. Overall outcome desired in managing an incident.

strike team. Unit formed through the combination of a number of like resources, such as brush trucks.

tactic. The deployment of resources in support of tactical objectives.

tactical objective. Operations that must be successfully performed if the strategic goals established for an incident are to be realized.

tanker shuttle. Use of a number of water tenders to transfer water from its source to the incident scene.

task force. A unit formed through the combination of different types of resources, such as several brush trucks and a bulldozer.

technical rescue. Utilization of specialized personnel and equipment in such challenging situations as confined space, high-angle, structural collapse, trench, and water rescues.

thermal column. Column of heated air, gases, and smoke that through convection rises above a fire.

toxic atmosphere. Atmosphere in which the air is potentially harmful.

traffic control. Control of vehicles on an incident scene in the interest of ensuring responder safety.

trench. Temporary excavation of a size and depth that presents the danger of a collapse trapping occupants.

trenching. Ventilation technique where a trench is cut in the roof of a building to allow the fire and heat to vent through the resulting opening.

undeclared emergency. Aircraft emergency where authorities on the ground do not receive advance notification.

under control. Term commonly used to indicate that an emergency situation has been brought under control or resolved.

unified command. Command structure where more than one individual commands the incident.

urban search and rescue (USAR). Search and rescue operations related to structural collapse and other technical and specialized rescue.

urban/wildland interface. Area where developed urban areas meet undeveloped wildland areas.

utilities. Infrastructure and services associated with the provision of electric, gas, sewer, telephone, and/or water.

vapor suppression. Measures taken to reduce the emission of vapors at a hazardous materials incident.

ventilation. Process of removing heated air, gases, and smoke from a building.

vertical ventilation. Ventilation technique involving the use of existing openings or cutting new openings in the roof of a building.

victim. Individual who suffers harm as a consequence of the occurrence of an emergency situation.

water-reactive materials. Materials that react when exposed to water.

water supply. Source of water used in firefighting.

working fire. Term commonly used to indicate a major fire.

Appendix B
Emergency Service Acronyms

AHJ	authority having jurisdiction
ALS	advanced life support
ATF	Bureau of Alcohol, Tobacco, Firearms, and Explosives
BLEVE	boiling liquid expanding vapor explosion
BLS	basic life support
CAFS	compressed air foam system
CISD	critical incident stress debriefing
CO	carbon monoxide
CP	command post
DHS	Department of Homeland Security
DOT	Department of Transportation
EMS	emergency medical services
EMT	emergency medical technician
EOC	emergency operations center
EOP	emergency operations plan
EPA	Environmental Protection Agency
ERG	Emergency Response Guidebook
FAA	Federal Aviation Administration
FEMA	Federal Emergency Management Agency
HSO	health and safety officer
IAP	incident action plan
IC	incident commander
ICP	incident command post
ICS	incident command system

IDLH	immediately dangerous to life and health
IMS	incident management system
IMT	incident management team
ISO	incident safety officer
JIC	joint information center
JIS	joint information system
LODD	line-of-duty death
MAC	multiagency coordination
MSDS	material safety data sheet
MVA	motor vehicle accident
NFA	National Fire Academy
NFFF	National Fallen Firefighters Foundation
NFIRS	National Fire Incident Reporting System
NFPA	National Fire Protection Association
NIMS	National Incident Management System
NIOSH	National Institute for Occupational Safety and Health
NTSB	National Transportation Safety Board
OSHA	Occupational Safety and Health Administration
PAR	personnel accountability report
PASS	personal alert safety system
PIO	public information officer
PPE	personal protective equipment
RIC	rapid intervention crew
RIT	rapid intervention ream
SCBA	self-contained breathing apparatus
SOG	standard operating guideline
SOP	standard operating procedure
USAR	urban search and rescue
USCG	United States Coast Guard
USFA	United States Fire Administration
VRT	vehicle rescue technician

Appendix C
Emergency Service Positions

The work of a fire department, both on and off the scene of emergency incidents, is performed by a cadre of dedicated professionals who enact the various roles and responsibilities of the positions they hold within the organization. These positions are categorized based on responsibility and rank within the fire department organizational structure and/or within the National Incident Management System (NIMS). Positions that are designated within the NIMS appear in italics below.

accountability officer. Individual assigned to track the location and activities of personnel operating on an incident scene.

assistant chief. Chief officer rank within a fire department.

battalion chief. Chief officer responsible for a number of companies or stations within a designated geographic area.

captain. Company officer rank within a fire department.

chief officer. Officer holding the rank of battalion chief, district chief, or higher.

command staff. The three staff positions within ICS responsible for supporting the incident commander as public information officer, safety officer, or liaison officer.

company officer. Officer responsible for a company within a fire department.

deputy chief. Chief officer rank within a fire department.

dispatcher. Staff member of an emergency communications center responsible for dispatching and coordinating radio communications with fire and emergency service units during an emergency incident.

district chief. Chief officer responsible for a number of companies or stations within a designated geographic area.

driver/operator. Individual responsible for driving fire apparatus to and from an incident, and for operating the apparatus during the emergency incident.

emergency medical technician (EMT). Emergency medical services provider trained in basic life support (BLS).

finance/administration section chief. General staff position within ICS responsible for managing all financial aspects of an incident.

fire chief. Senior ranking chief officer within a fire department.

firefighter. Member of a fire department who does not hold an officer position or rank.

fire inspector. Individual responsible for conducting inspections and enforcing building and fire codes.

fire investigator. Individual responsible for determining the origin and cause of a fire.

fire police. Fire department member responsible for traffic and scene control at an emergency incident.

general staff. The four positions within ICS responsible for each of the functional sections: operations, planning, logistics, and finance/administration.

health and safety officer (HSO). Fire department member responsible for managing a fire department's health and safety program.

incident commander (IC). Individual designated within ICS with responsibility for the overall management of an emergency incident.

incident safety officer (ISO). Command staff position within ICS responsible for ensuring the safety of personnel operating on an incident scene (also called *safety officer*).

landing zone officer. Individual assigned responsibility for coordinating the landing of a helicopter.

liaison officer. Command staff position within ICS responsible for coordinating with representatives of cooperating and assisting agencies.

lieutenant. Company officer rank within a fire department.

logistics section chief. General staff position within ICS responsible for ensuring that the support and service needs of an incident are addressed.

officer. Fire department member responsible for the supervision of other personnel.

operations section chief. General staff position within ICS responsible for managing all tactical operations at an incident.

paramedic. Emergency medical services provider trained in advanced life support (ALS).

planning section chief. General staff position within ICS responsible for planning activities at an incident.

public information officer (PIO). Command staff position within ICS responsible for working with the media and the public in the interest of ensuring the accuracy of information disseminated.

rescue officer. Officer assigned to a rescue company or in charge of a particular rescue operation at an emergency incident.

sergeant. Company officer rank within a fire department.

technical specialist. Individual with advanced training and certification in a specialization, such as hazardous materials or technical rescue.

wildland firefighter. Firefighter with specialized training in wildland firefighting.

Appendix D
Emergency Service Apparatus

Vehicles designed for firefighting and other emergency service applications are referred to as *apparatus*. Apparatus can be categorized by purpose and type.

aerial apparatus. Apparatus equipped with hydraulically operated ladders or elevated platforms are classified as aerial apparatus. Aerial apparatus enable firefighters to gain access to elevated positions, facilitate rescues, and apply large quantities of water to the upper floors of buildings. Aerial apparatus incorporate various aerial devices based on their intended purpose, with firefighting and rescue operations being supported by aerial ladders, elevating platforms, and ladder pipes. In addition to a main ladder, typically ranging from 75 to 100 feet in length, aerial apparatus carry a complement of ground ladders. Aerial apparatus are commonly referred to as *trucks*.

aircraft. Both fixed-wing and rotary aircraft are utilized in firefighting and rescue operations. These specialized aircraft are designed in accordance with their purpose. Helicopters can incorporate rescue and patient transport capabilities. Both fixed-wing planes and helicopters are used in the transport and application of water and other extinguishing agents in wildland firefighting.

combination apparatus. The functionality of apparatus is often enhanced through the incorporation of more than one traditional feature or set of capabilities into a single apparatus, as in the case of a pumper/tanker, rescue pumper, or paramedic engine. Traditional combinations that are commonplace within the fire service include quads and quints. A *quad* combines the water tank, pump, and hose load of an *engine*, with the ladders typically carried on a *truck*. The addition of an aerial device enhances the capabilities of a quad to what is commonly called a "quint."

emergency medical units. Emergency medical service units provide for the ground or air transportation of patients, with air transportation usually being provided by helicopters. Emergency medical service units are classified in accordance with their capabilities and include quick response (QRS) units, basic life support (BLS) units, and advanced life support (ALS) units.

firefighting apparatus. Firefighting apparatus is designed to support fire attack and related operations. The basic firefighting apparatus incorporates hose, a water supply, and pumping capacity and is referred to as an *engine* or *pumper*. Fire department pumpers are further classified based on pumping capacity and amount of water carried.

rescue apparatus. Apparatus designed for use in various types of rescue operations is considered rescue apparatus. Rescue units are categorized based on their capabilities and the equipment that they carry, ranging from light rescue vehicles capable of handling basic extrications, to medium rescue vehicles that carry an expanded complement of power extrication and other rescue equipment, to heavy rescue vehicles that carry the extensive tools and equipment required in specialized and technical rescue operations.

reserve apparatus. Reserve apparatus includes apparatus that, while no longer in service, is in full operational condition and available to be placed into service when needed, as in the case when a major incident or multiple simultaneous incidents result in a shortage of apparatus to respond to additional calls.

specialized apparatus. Specialized apparatus is designed to meet the requirements of particular firefighting operations. The brush trucks, designed for use in wildland firefighting, are supported by traditional heavy equipment, including bulldozers. Specially designed and equipped apparatus, such as crash rescue trucks, are utilized in aircraft rescue and fire fighting (ARFF). Fireboats and marine units are designed for use in firefighting and rescue operations.

support apparatus. Support apparatus are designed to provide for the service and support needs of an emergency incident. Air supply units equipped with cascade systems provide for the on-scene replenishment of breathing air in self-contained breathing apparatus (SCBA). Communication units provide enhanced communications capabilities at an incident scene. Canteen units service the needs of personnel operating on an incident scene.

water supply apparatus. The purpose of water supply apparatus is to provide for the transport of large quantities of water to a fire scene. Mobile water supply apparatus have capacities that typically range from 3,000 to 6,000 gallons and are commonly referred to as *tankers* or *tenders*. Under the resource typing system implemented throughout the fire service, the proper nomenclature for ground-based mobile water supply units is *tenders*, with aircraft water supply apparatus being considered *tankers*. While water is the traditional extinguishing agent utilized in firefighting, various types of foam concentrates have been found to be effective in particular firefighting operations. Many fire apparatus now incorporate compressed air foam systems (CAFS) that in addition to combining water and foam concentrate, also incorporate air, producing an extinguishing agent with enhanced capabilities. Apparatus that incorporate foam capabilities include foam pumpers, foam tankers, and foam trailers.

Appendix E
Emergency Service Equipment

Portable tools, appliances, and other equipment carried on fire apparatus that are removable from the apparatus are referred to as "equipment." While some equipment is applicable to many types of incidents, other more specialized equipment is designed for use at specific types of incidents. For example, a portable generator might be used to power equipment and provide lighting at a variety of incidents. Specialized rescue equipment, including power extrication tools, would primarily be used at incidents involving vehicle, technical, or specialized rescues.

air bank (cascade system). Set of large cylinders containing breathing air used in refilling SCBA bottles.

air pack. Respiratory device worn by emergency responders that provides breathing air from an air cylinder (also called a self-contained breathing apparatus (SCBA).

air bag. Inflatable bag used during rescue operations to lift or separate objects.

attack line. Hoseline supplied by a fire pumper that firefighters use in fire suppression activities.

chemical foam. Extinguishing agent produced through the combination of foam concentrate and water.

cribbing. Lumber precut to various lengths used to stabilize vehicles and buildings during various types of rescue operations.

cutting tool. Hand or power tool used to cut various materials.

deck gun. Master stream appliance designed to deliver large quantities of water.

extinguishing agent. Substance or material used in the control or extinguishment of a fire.

fire extinguisher. Portable extinguishing device designed to fight a fire in its early stages.

gas detector. Electronic device designed to detect the presence of various gases.

generator. Gasoline- or diesel-powered device used to produce electrical power.

ground ladder. Portable ladder, available in various lengths and types, designed for use by firefighters in laddering a building.

hand tool. Tool that derives its energy from that of the person using it.

handline. A hoseline used by firefighters in fire suppression activities.

high-rise pack. A preassembled collection of fire hose, a nozzle, adapters, and other tools used in firefighting in high-rise buildings.

hydraulic jack. A tool that uses hydraulic power for lifting.

hydraulic shoring. Shoring equipment used in rescue operations that uses hydraulic power.

life belt. Adjustable belt with an attached device for securing a firefighter to a ladder.

lifeline rope. Specialized rope designed to ensure life safety during rescue operations.

life safety harness. Personal protective device worn by emergency responders during certain technical rescue operations.

nozzle. Appliance that when attached to the end of a handline provides for control of the volume and pattern of water application.

personal alert safety system (PASS). Motion sensing device worn by firefighters that ensures their safety by alerting after a designated period of lack of movement.

personal protective equipment (PPE). Ensemble of protective clothing worn by emergency responders that includes a coat, pants, helmet, boots, gloves, eye protection, a protective hood, a self-contained breathing apparatus, and a personal alert safety system device.

pneumatic shoring. Shoring equipment used in certain rescue operations that is powered by compressed gas.

pneumatic tool. Tool powered by compressed gas.

portable equipment. Removable tools and equipment carried on fire apparatus.

positive pressure ventilation fan. Gas- or electric-powered fan used to blow fresh air into a confined space or building during positive pressure ventilation (PPV).

power extrication tools. Extrication tools, including spreaders, shears, and rams, powered by hydraulic power units.

power tool. Tool powered by a motor or pump.

proximity clothing. Specialized personal protective equipment designed to protect firefighters as they work in close proximity to a fire.

prying tool. Hand tool designed to be used as a lever during firefighting and rescue operations.

rescue rope. Specialized rope reserved for use during rescue operations involving the lowering or raising of personnel.

rigging. Ropes and other devices used in lifting operations.

self-contained breathing apparatus (SCBA). Respiratory device worn by emergency responders that provides breathing air from an air cylinder (also called an airpack).

shoring. Equipment and materials used to provide temporary support and stabilization during technical rescue operations.

smoke ejector. Gasoline or electrical powered device used to remove smoke from a building or blow fresh air into a building or confined space.

special protective clothing. Protective clothing designed to protect emergency responders during hazardous materials incidents.

Stokes basket. Device incorporating a litter and protective basket used in raising, lowering, or transporting victims.

striking tool. Tool that incorporates a weighted head and a handle, such as an axe or sledgehammer.

structural turnout gear. Personal protective equipment (PPE) designed for use in structural firefighting, which incorporates the essential components listed above under "personal protective equipment."

supply hose. A large diameter hose (LDH) used to transport a large volume of water from a water source, such as a fire hydrant or drafting source, to the fire pumper supplying handlines and/or master streams on an incident scene.

thermal imaging camera (TIC). A device designed to detect sources of heat and enhance the vision of firefighters operating in low-visibility environments.

Appendix F

Governmental Agencies and Private Organizations

Bureau of Alcohol, Tobacco, Firearms, and Explosives (ATF). The Bureau of Alcohol, Tobacco, Firearms, and Explosives is a federal law enforcement agency within the U.S. Department of Justice. Its responsibilities include the investigation of acts of arson and bombings.

Department of Homeland Security (DHS). The U.S. Department of Homeland Security is a federal Cabinet department that was established on November 25, 2002, in response to the September 11 attacks. The major government reorganization that established the Department of Homeland Security incorporated 22 formerly separate governmental agencies or units. The Department has responsibilities for preventing and responding to terrorist attacks, man-made accidents, and natural disasters.

Department of Transportation (DOT). The U.S. Department of Transportation is a federal Cabinet department. Its responsibilities include the regulation and enforcement of the national transportation system and services.

Environmental Protection Agency (EPA). The U.S. Environmental Protection Agency is charged with the role of protecting the health of humans and the environment. Its responsibilities include developing, issuing, and enforcing related regulations.

Federal Aviation Administration (FAA). The Federal Aviation Administration, an agency within the U.S. Department of Transportation, is the designated aviation authority. It is responsible for the regulation and oversight of all civilian aviation within the United States.

Federal Emergency Management Agency (FEMA). The Federal Emergency Management Agency is an agency within the U.S. Department of Homeland Security. Its responsibilities include the coordination of the response to disasters in those instances where resource needs exceed the capabilities of local and state resources.

National Fallen Firefighters Foundation (NFFF). The National Fallen Firefighters Foundation is a nonprofit organization with missions of honoring firefighters who have sacrificed their lives in the line of duty and reducing firefighter fatalities. It offers training programs and provides services to surviving families and fire departments.

National Fire Academy (NFA). The National Fire Academy is a training and education institution within the U.S. Fire Administration. It provides management and technical training for fire and emergency services personnel.

National Fire Protection Association (NFPA). The National Fire Protection Association is a private organization with a mission of the development of standards for fire prevention, protection, and suppression. The voluntary consensus standards developed by this standards-making group address all aspects of a fire department's operations, including the qualifications of personnel, standards for fire apparatus and equipment, and incident management.

National Institute for Occupational Safety and Health (NIOSH). The National Institute for Occupational Safety and Health is an agency within the U.S. Department of Health and Human Services. Its responsibilities for conducting research and developing recommendations to reduce work-related injuries, illnesses, and fatalities include the investigation of firefighter line-of-duty deaths.

National Transportation Safety Board (NTSB). The National Transportation Safety Board is an independent federal investigative agency. In addition to its responsibility for investigating aviation incidents and accidents, the NTSB also investigates certain highway, railroad, and marine accidents, as well as pipeline incidents.

Occupational Safety and Health Administration (OSHA). The Occupational Safety and Health Administration is an agency within the U.S. Department of Labor. Its responsibilities include preventing workplace injuries, illnesses, and fatalities through the issuance and enforcement of workplace safety regulations.

U.S. Coast Guard (USCG). The U.S. Coast Guard is a branch of the U.S. Armed Services, which during peacetime operates under the authority of the U.S. Department of Homeland Security. Its responsibilities include the enforcement of maritime regulations.

U.S. Fire Administration (USFA). The U.S. Fire Administration is a division of the Federal Emergency Management Agency. Its primary responsibilities are related to reducing the incidence of fire within the United States and include conducting relevant research and developing and delivering training courses and educational programs for fire and emergency service professionals through the National Fire Academy (NFA) and similar programs for emergency management personnel through its sister organization, the Emergency Management Institute (EMI).

Appendix G

Public Information Officer Sample Position Description

Position Description

The public information officer (PIO) coordinates the gathering and dissemination of operational and administrative information regarding the fire department to the public and the media. Responsibilities of the position include serving as the fire department's PIO at emergency incidents and handling routine public information activities off the incident scene. Major components of this position involve public information, public relations, and public education. The PIO arranges and conducts media interviews, prepares and issues press releases and media advisories, and responds to inquiries from the public, the media, and governmental agencies and officials.

Major Duties and Responsibilities

The PIO enacts the following responsibilities, based on *NFPA 1035, Standard for Professional Qualifications for Fire and Life Safety Educator, Public Information Officer, and Juvenile Firesetter Intervention Specialist* (2010).

1. Conducts media interviews at emergency incidents and other events in accordance with established fire department policies and procedures.
2. Coordinates dissemination of information to designated groups, organizations, and agencies, including the public.
3. Prepares and disseminates press releases to the media regarding emergency operations and other fire department activities and programs.

4. Coordinates press conferences for local media related to fire department operations, activities, and programs.
5. Serves as PIO within the incident management system at major emergency incidents and other incidents of public and media interest.
6. Coordinates with appropriate fire department personnel, including the incident commander on an incident scene, in determining the appropriate nature and manner of the release of information.
7. Maintains current understanding of all fire department operations and activities.
8. Develops and maintains appropriate working relationships with fire and emergency service operational and administrative personnel, the public and community groups, and the media.
9. Responds in an accurate and timely manner to questions from the public and media regarding the fire department, its operations, and activities.
10. Performs other duties as assigned.

Reporting Relationships

The PIO reports to the fire chief.

Supervisory Relationships

The PIO is responsible for overseeing the work of fire department personnel assigned to public information activities during the performance of these duties.

Working Relationships

The PIO is expected to interact professionally with the public and community groups, the media, elected and appointed governmental officials, and members of the fire and emergency services.

Working Conditions

The PIO is expected to enact the above responsibilities as needed based on the public information needs of the fire department and the community that it serves. This will often require varied working hours that may include evenings and weekends.

Work Environment

The PIO will perform the above activities both on and off the incident scene, including at the fire department's stations.

Appendix H

National Incident Management System Overview

The National Incident Management System (NIMS), implemented by the U.S. Department of Homeland Security, is designed to ensure the effective, efficient, and safe management of emergency incidents in accordance with recognized incident management priorities. It provides a robust incident management tool that is applicable regardless of the type of incident and provides for the development of an appropriate organizational structure tailored to the specific management needs of a given emergency incident (fig. A–1).[1]

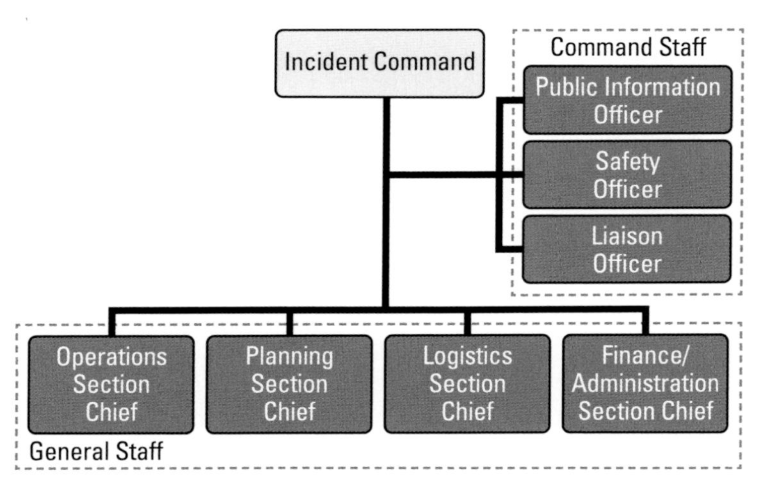

Fig. A–1. Incident command system organization chart

The person designated with the overall responsibility for managing a particular incident is referred to as the incident commander (IC). Most incidents will involve a *single command* wherein an individual, such as the fire chief or a senior officer, serves as the incident commander.

In a *unified command* more than one individual enacts the command functions that contribute to successful incident management.

The use of NIMS in incident management was discussed in chapter 8. The distinction between command staff positions that perform staff activities in support of the incident commander and general staff positions responsible for functional aspects of the management of an emergency incident was likewise considered. In designing an appropriate organizational structure that is responsive to the management needs of the incident, an incident commander will establish necessary command and general staff positions and appoint qualified individuals to staff each position.

The roles and responsibilities of the incident commander were discussed in chapter 8, as were the associated roles and responsibilities of the public information officer, a command staff position within the Incident Command System incorporated within the National Incident Management System. This appendix will provide an overview of the roles and responsibilities of the two additional command staff positions, safety officer and liaison officer, and those of the four general staff positions of the section chiefs of the operations, planning, logistics, and finance/administration sections.

Public Information Officer

The public information officer (PIO) will at times be expected to enact certain responsibilities in addition to those discussed in chapter 8. This is often the case when the nature, size, and/or scope of the emergency incident(s) result in the establishment of either an area command or a multiagency coordination system. An *area command,* also called a unified area command, is an organizational structure designed to oversee the management of multiple incidents that are each being handled by a separate ICS organization, or large or multiple incidents to which several incident management teams have been assigned. The area command has the responsibility for determining overall strategy and priorities, allocating critical resources in accordance with priorities, and ensuring that each incident in properly managed. The utilization of a multiagency coordination system (MACS) likewise provides for the coordination of the resources of assisting agencies to support emergency operations in accordance with established priorities.

A joint information system (JIS) will often be utilized in such situations in the interest of providing consistent, coordinated, and timely information regarding emergency incidents. The mission of a JIS is to provide a structure, in the form of a joint information center (JIC), and system for developing and disseminating coordinated interagency messages, developing and implementing public information plans and strategies on the incident commander's behalf, bringing relevant public affairs issues to the attention of the incident commander, and controlling inaccurate information and rumors that could undermine the public's confidence regarding how an emergency incident is being managed. The role and responsibilities of a PIO will thus be different in those instances where the coordination of information dissemination has been deemed important and a JIC and JIS are providing this desired coordination.

Safety Officer

The safety officer is a member of the command staff responsible for monitoring and assessing safety hazards and unsafe conditions and developing measures for ensuring the safety of personnel operating on an emergency incident scene. The following are the major responsibilities of the safety officer under NIMS:

- Identify and mitigate hazardous situations.
- Ensure safety messages and briefings are made.
- Exercise emergency authority to stop and prevent unsafe acts.
- Review the incident action plan for safety implications.
- Assign assistants qualified to evaluate special hazards.
- Initiate preliminary investigation of accidents within the incident area.
- Review and approve the medical plan.
- Participate in planning meetings.

Liaison Officer

The liaison officer is a member of the command staff and is responsible for the coordination of incident activities with the various representatives from cooperating and assisting agencies. The following are the major responsibilities of the liaison officer under NIMS:

- Act as a point of contact for agency representatives.
- Maintain a list of assisting and cooperating agencies and agency representatives.
- Assist in setting up and coordinating interagency contacts.
- Monitor incident operations to identify current or potential interorganizational problems.
- Participate in planning meetings, providing current resource status, including limitations and capabilities of agency resources.
- Provide agency-specific demobilization information and requirements.

Operations Section Chief

The operations section chief is responsible for the management of all tactical operations at an incident. The following are the major responsibilities of the operations section chief under NIMS:

- Ensure safety of tactical operations.
- Manage tactical operations.
- Develop the operations portions of the incident action plan.
- Supervise execution of the operations portions of the incident action plan.
- Request additional resources to support tactical operations.
- Approve release of resources from active operational assignments.
- Make or approve expedient changes to the incident action plan.
- Maintain close contact with the incident commander, operations personnel, and other agencies involved in the incident.

Planning Section Chief

The planning section chief is responsible for the collection and dissemination of information used in incident planning activities. The following are the major responsibilities of the planning section chief under NIMS:

- Collect and manage all incident-related operational data.
- Supervise preparation of the incident action plan.
- Provide input to the incident commander and Operations in preparing the incident action plan.
- Incorporate traffic, medical, and communications plans and other supporting material into the incident action plan.
- Conduct and facilitate planning meetings.
- Reassign personnel within the ICS organization.
- Compile and display incident status information.
- Establish information requirements and reporting schedules for the resources and situation units.
- Determine need for specialized resources.
- Assemble and disassemble task forces and strike teams not assigned to operations.
- Establish specialized data collection systems as necessary.
- Assemble information on alternative strategies.
- Provide periodic predictions on incident potential.
- Report significant changes in incident status.
- Oversee preparation of the demobilization plan.

Logistics Section Chief

The logistics section chief is responsible for ensuring that necessary support and services are provided at an emergency incident. The following are the major responsibilities of the logistics section chief under NIMS:

- Provide all facilities, transportation, communications, supplies, equipment, maintenance and fueling, food and medical services for incident personnel, and all off-incident resources.
- Manage all incident logistics.
- Provide logistical input to the incident action plan.
- Brief Logistics staff as needed.
- Identify anticipated and known incident service and support requirements.
- Request additional resources as needed.
- Ensure and oversee the development of the communications, medical, and traffic plans as required.
- Oversee demobilization of the Logistics section and associated resources.

Finance/Administration Section Chief

The finance/administration section chief is responsible for the management of all financial aspects of an incident. The following are the major responsibilities of the finance/administration section chief under NIMS:

- Manage all financial aspects of an incident.
- Provide financial and cost analysis information as requested.
- Ensure compensation and claims functions are being addressed relative to the incident.
- Gather pertinent information from briefings with responsible agencies.

- Develop an operating plan for the finance/administration section and fill section supply and support needs.
- Determine the need to set up and operate an incident commissary.
- Meet with assisting and cooperating agency representatives as needed.
- Maintain daily contact with agency(s) headquarters on financial matters.
- Ensure that personnel time records are completed accurately and transmitted to home agencies.
- Ensure that all obligation documents initiated at the incident are properly prepared and completed.
- Brief agency administrative personnel on all incident-related financial issues needing attention or follow-up.
- Provide input to the incident action plan.

Further information on NIMS is available through the website of the Federal Emergency Management Agency (FEMA) (www.fema.gov/emergency/nims).

Notes

1 U.S. Department of Homeland Security. 2008. *National Incident Management System*. Emmitsburg, MD: U.S. Department of Homeland Security. December.

Index

Glossaries are provided in appendices starting on page 365: Terminology, Acronyms, Positions, Apparatus, Equipment, and Government Agencies and Organizations.

A

absorption, 224
accessibility, 82
accident investigating, 131
accident victims, 128
accidental fires, 344
accidents
apparatus dispatched for, 131
causes of, 131
job aids, 134–136
media coverage of, 130–133
operational overview, 127–129
priorities, 127
resource requirements, 130–131
safety considerations, 129–130
typical incidents, 125–126
acronyms, 106, 121
activity, of early minutes, 19
aerial apparatus, 258
After the Fire! Returning to Normal (USFA), 120
agricultural rescue, 251
air bags, 209
air quality monitoring, 161
aircraft accidents, 131, 355
aircraft surveillance, 271
aircraft vehicles, 126. *See also* helicopters
air-drops, 191, 193
air-monitoring, 175
air-monitoring equipment, 174
airplane crash, 353
Alfred P. Murrah Federal Building bombing, 284
alternate fuels, 209
animal rescue, 253
apparatus and equipment. *See also* self-contained breathing apparatus (SCBA)
aerial, 210, 258
firefighting, 176, 210, 225
incident need determines, 131
needs for, 20, 121, 146, 162, 176–177, 194, 225, 241, 291, 309
rescue, 131, 177, 210, 225, 257
specialized, 225, 257, 296
staging, 174, 176, 222, 238
water supply, 177, 194, 210, 225
apparatus and personnel staging, 308
apparatus dispatched
for accidents, 131
for building fires, 146
for infrastructure emergencies, 241
for vehicle fires, 210
for wildland fires, 194
Are You Ready? A Guide to Citizen Preparedness (FEMA), 311
arson and incendiary fires
about, 341
classifying the cause of a fire, 344
collecting and preserving evidence, 344
determining the origin and causes of fire, 343
fire service roles and fire

investigation, 342
media coverage job aids, 347–348
media coverage of arson and incendiary fires, 345–346
arson investigators, 343
asphyxiation, 286
assignment editors, 67
authority having jurisdiction (AHJ), 140

B

backburning, 191
background research, 74
backup emergency generators, 235
beat assignment reporters, 71
biological events, 285
blogs, 93–94
body recovery, 268
boiler fires, 157
boiling liquid expanding vapor explosion (BLEVE), 176, 210
bomb robots, 290
bombings, 284
breaking news stories
coverage of, 32
major, 69
timely dissemination, 80
broadcast professionals, 70
brush trucks, 191
building collapse, 258
building evacuation, 142
building fires
about, 139
apparatus dispatched, 146
incident priorities, 142
job aids, incident management of, 149–150
job aids, media coverage of, 150–153
media coverage of, 147–150
needs of, 320–321
resource requirements, 146
safety considerations, 145–146
statistics, 139
types of, 143
typical incidents, 140–142
typical tools, 146
building utilities, 144
building/structural integrity, 118
Bureau of Alcohol, Tobacco, Firearms and Explosives (ATF), 283, 291
byline, 66

C

camera phones, 23
canine search dogs, 271
carbon monoxide incidents, 219
cargo placards, 210
cave-in, 267
chemical burns, 286
chemical events, 285
Civilian Fire Fatalities in Residential Buildings (USFA), 320
civilian injuries and fatalities
about, 319–320
media coverage job aids, 326–327
media coverage of, 322–325
stakeholder expectations, 320–322
cleaning materials, 170
cold zone, 111, 174, 221
command positions, 114–115
command staff, 51
commercial processes, 170
commitments
proactive, 47
in working relationship, 81, 84
communication
capability of, 273
patterns, 101
preparation for, 82
proactive, 88
proactive vs. reactive strategies, 46
problems, 193
systems, 307
communication skills, 56, 74
communication tools, 24
communication units, 275
community response plans, 220
compactor files, 157
composure, 85
compressed natural gas, 205
confidential information, 82

confined space rescue, 252
confinement, 175
control management, 240
control zones
  crime scene, 290
  crowd and traffic control, 208, 237, 240, 274
  hazard zones, 129
  hazardous materials, 224
  infrastructure emergencies, 241
  management of, 357
  protective, 174, 221
  use of, 111
  vehicle fires, 209
cooling, 176
coverage
  of early minutes, 19
  of high-profile incidents, 7
  related, 48
  sensitivity of, 319
coverage assignment, 66–67
coverage decisions, 12–13
coverage opportunities, 42
coverage sharing, 355
credibility, 57
crew safety, 193
crime scene, 289–290
criminal investigation, 293
crowd control, 175, 237, 274, 289
crown fires, 193
currency, 58

D

daily newspapers, 36
deadlines, 35–37, 81
debriefing, 130
decisional roles, 48
decontamination, 162, 176, 223
defensive attack mode, 117, 144–145
defensive strategy, 222
Department of Homeland Security (DHS), 291
derailment, 353
detours, 178, 194
digital camera, 57
digital camera optics and technology, 98
digital images, 94
dispatch, 107
dispatcher, 116
dissemination, 69
disseminator roles, 48–49, 56
dive rescue teams, 257
drownings, 267–268

E

earthquakes, 92, 302, 308, 352
editors, 4
educational agenda, 28
effectiveness, 18
electrical circuit de-energization, 159, 162
electrical emergencies, 235
electrical equipment fires, 157
electrical power lines
  downed, 234–235, 238, 241, 250, 300, 305–306
  fire sources, 188
  medical helicopters, 130–131
  wildland fires, 193
electrocution, 161
elementary school, 141
elevators, 235
emergency incident coverage
  media expectations perspective, 32–33
  success in, 75
emergency incident coverage, fire department perspective
  about, 17–18
  action bias, 18–19
  community risk reduction, 28
  getting information to the public, 25–26
  incident management initial focus, 21–22
  incident management process and priorities, 19–20
  incident reporting, 23–25
  reporting and related stories, 26–27
  success through collaboration, 28–29
emergency incident coverage, media perspective

about, 31–32
bad days, 42
media perceived, 43
news coverage essence, 32–33
news media expectations, 32–33
news media expectations variability, 35–40
reporters needs in incident coverage, 41–42
technology and news reporting, 40

emergency incidents
beginnings of, 5–6
change, 7
covering of, 13–16
under-informed reporting of, 10–11
media coverage of, 11, 359
news coverage and technology, 7–8
news coverage of, 4
news coverage roles, 66
non-routine nature of, 6–7
preparation for, 11–12
scene of, 71
specialists vs. generalists, 12–13
success incident covering, 8–10
successful coverage of, 51

emergency incidents coverage in technology age
about, 91–92
changing paradigms in news reporting, 96–97
changing players, 97–99
Internet and social media, 92–96
Internet and social media in breaking news stories, 100–103

emergency management agencies
jurisdiction, 302
local, 291, 309
and response agencies, 276

Emergency Management Institute (EMI), 59

emergency medical service units, 258, 289

emergency responders
life safety of, 189, 253
rehabilitation, 223

Emergency Response Guidebook (U.S. Department of Transportation), 220

emergency service organizations, expectations of, 4

environmental factors, 174

Environmental Protection Agency (EPA), 355

equipment fires
about, 155–156
categories of, 157
electrocution, 161
exposure protection, 161
firefighting tactics, 161
heating and cooking fires, 157
ignition factors, 155
impact of, 163
incident management job aids, 166
incident priorities, 158
incidents, typical, 156–158
job aids, 164
media coverage job aids, 166–167
media coverage of, 163–164
operational overview, 158–161
resource requirements, 162–163
safety considerations, 158–162
tools, typical, 162–163

escape routes, 193

etiological harm, 286

evacuation, 11, 111, 142, 221

evacuations, 174, 178, 240–241

Everyone Goes Home program, 331

explosions, 188

explosive events, 286

exposure protection, 158, 161, 175, 208, 223

exposures, 142, 238

extinguishing agent, 160, 162, 175

F

Facebook, 8, 94, 99

fall prevention, 28

fatalities. *See also* injuries and fatalities
accurate reporting on, 23
official sources on, 148
residential buildings, 320

Federal Bureau of Investigation (FBI), 283–284, 291

Federal Emergency Management Agency (FEMA), 6, 112, 291, 311

federal government agencies, 121, 132
field coverage enhancement, 68
field reporters, 70–71, 73
final thoughts, 359–362
fire and emergency services relationship with media, 1
fire burn patterns, 343
fire cause dissemination, 345–346
fire chief, 20, 49, 113, 118, 345
fire confinement, 158
fire department
  evolution of, 14
  expectations of, 4
  media coverage response, 79
  role and services, 14
  staffing, 4
fire department and the media representatives, 243
fire department connection piping, 145
fire department representatives, 8
fire department websites, 99
fire department working with the media
  about, 77–78
  before an emergency incident, 85–86
  during an emergency incident, 86–87
  after an emergency incident, 87–88
  principles of, 82–84
  successful media coverage responsibility, 78–79
  working relationship, 78–81
Fire Engineering (PennWell/FireEngineering), 122
fire hydrants, 143
fire intensity, 193
fire investigation, 341
fire investigators, 210, 342–343
fire marshal, 345
fire prevention, 26
fire service books, 122
fire spread
  building fires, 142
  exposure protection, 144
  outside fires, 195
  rate of, 193
  wildland interface, 188, 190, 193
fire-control lines, 191
firefighter injuries and fatalities
  about, 329–331
  incident within incident, 331–332
  media coverage job aids, 337–339
  media coverage of, 333–336
  priorities, 329
firefighter line-of-duty death (LODD), 331
firefighter rescue, 253
firefighting decisions, 11
firefighting tactics, 161, 175
Fire-Related Firefighter Injuries Reported to NFIRS (USAF), 330
fires intentionally set, 341
fireworks, 188
five pillars of journalism, 32
flawed operations, 24
flooding, 242–243, 305, 308
follow up
  coverage, 147, 345
  information, 88
forecasting, 221
framing the story, 87
freelancing, 84, 193
fuel load, 193
fuel supplies, 162
fuel tank, 170
funeral of firefighter(s), 335

G

gas emergency, 242
gas leaks, 27, 224
gas line fire, 176
gas mains
  and distribution equipment, 236
  leaks, 306
  rupture of, 170
  security of, 174
gas pipelines, 171
gas supply, 174
gas-line ruptures, misreporting of, 10
gasoline, 172
general assignment reporters, 71
general staff assignment, 114

golden hour, 21–22
governmental agencies, 113, 130–131, 176

H

handlines, 144
harm, types of, 286
hashtags, 95
hazard classification system, 219
hazard control zones, 129
hazardous materials
    building fires, 142
    complications, 129
    misreporting of leaks, 10–11
    nature of, 219
    as power sources, 156
    properties of, 160–162
    in vehicle fires, 205, 208–209
hazardous materials fires
    about, 169–170
    explosive potential, 172
    firefighting tactics, 175
    incident management job aids, 179–180, 197–198
    incident priorities, 172
    incidents, typical, 170–172
    media coverage job aids, 180–183, 198–201
    media coverage of, 177–178
    operational overview, 172–175
    reactivity basis of, 172
    resource requirements, 176–177
    safety considerations, 175–176
    size-up and forecasting, 177
    tools, typical, 177
hazardous materials incidents
    about, 217
    incident management job aids, 227–278
    incident priorities, 221
    incidents, typical, 218–220
    media coverage job aids, 228–231
    media coverage of, 225–226
    operating principles, 222
    operational overview, 221–223
    operational tactics, 222–223
    resource requirements, 224–225
    safety considerations, 223–224
    tools, typical, 225
hazardous materials release, environmental impact of, 226
hazardous materials technicians, 224
hazards, inherent, 240
"hazmat" incidents. *See* hazardous materials incidents
heart attacks, 330
helicopters
    medical, 130–131
    news, 23, 37, 40, 67, 147, 211, 225
    search and rescue, 257–258, 275
    wildland fires, 194
high-angle rescues, 252, 259
high-profile incidents, coverage of, 7
high-rise buildings, 145
hospital fire plans, 141
hospital fires, 141
hot spots, 161
hot zone, 111, 129, 174, 221, 223–224
human factors
    equipment fires, 155
    vehicle fires, 205
human interest stories, 275, 310
Hurricane Irene, 92–93
hurricanes, 310, 352–353
hybrid vehicles, 129–130, 209
hydrant systems, 145
hydrogen (H) fuel, 205

I

ice rescues, 251
inaccuracies in reporting, 10, 80
incendiary fires, 344
incident action plan (IAP), 256
incident action plan (IAP) development, 118
Incident Command System (ICS), 19, 112
incident commander
    availability of, 22
    comprehensive size-up, 305
    damage assessment, 305
    on fire cause, 345
    fire investigation role, 342, 345
    firefighter emergency, 331–332, 334
    focus of, 51

high visibility incidents, 356–357
incident action plan (IAP), 118–119, 189, 256
incident management responsibilties, 113–115
information collection briefing, 61
information dissemination, 110
ISO assignment and role, 240, 256
media coverage interest, 8
media coverage of accidents, 131–133
medical surveillance, 223
mutual aid request, 6, 116
national public and media interest, 356
news media interviews, 147, 226, 259, 276
news release approval, 115
notifications be made to state and/or federal agencies, 355
operational overview, 127–128, 143–145
preplanning activities, 109
press interviews, 82
public information officer (PIO), 42, 52, 54
public information officer (PIO) assignment, 334
public information officer (PIO) designation, 360
public information officer (PIO) guidance and support from, 55
reporting inaccuracies, 24
responsibilities of, 363
role in incident of national public and media interest, 356
role of, 87
safety considerations, 145–146, 161
situation evaluation, 20
incident coverage monitoring, 101
incident guides, 15
incident management
about, 107–108
example, 20
incident action plan development, 118–120
incident commander management responsibilities, 113–115
incident management system, 111–113
learning more about, 121–122
priorities, 108–110
process of, 116–117
public information officer (PIO) responsibilities, 115
resources for emergency incident media coverage, 121
of wildland fires, 190
working with media coverage, 110–111
incident management job aids
building fires, 149–150
equipment fires, 166
hazardous materials fires, 179–180, 197–198
hazardous materials incidents, 227–278
infrastructure emergencies job aids, 244–245
job aids, 134–136
rescues job aids, 260–262
searches job aids, 277–279
terrorism incidents job aids, 294–296
vehicle accidents, 134
vehicle fires, 212–213
weather-related events and national emergencies job aids, 312–314
incident management priorities
incident stabilization, 108
life safety, 108
property conservation, 108
incident management process, 117
incident management system, 51
incident management systems, 111, 290
incident occurrences alerting public, 100
incident priorities
equipment fires, 158
hazardous materials incidents, 221
incident stabilization, 127, 142, 158, 172, 188, 206, 221, 237, 253, 270, 287, 304
life safety, 127, 142, 158, 172, 188, 206, 221, 237, 253, 270, 287, 304

operational overview, 143–145
outside fires, 188
property conservation, 127, 142, 158, 172, 188, 206, 221, 237, 253, 270, 287, 304
typical incidents, 237
vehicle fires, 206

incident safety officer (ISO). *See also* life safety; personnel accountability
appointment of, 111, 119, 145, 274
functions, 114
incident management priorities, 162, 307, 332
infrastructure emergencies, 240
media area coordination, 62
medical surveillance, 223
and rescue officer, 256
rescues, 256
role of, 175
safety considerations, 145–146
use of, 129, 290
vehicle fires, 210
wildland fires, 193

incident scene, access to, 42

incident stabilization, 19, 51, 110, 172, 188, 221, 237, 253, 270, 287, 304, 329

incident types, knowledge of, 41

incident-related injuries and fatalities, reporting on, 324

incidents, high profile and visibility, 292

incidents, typical
accidents, vehicular, 125–126
building fires, 140–142
equipment fires, 156–158
hazardous materials fires, 170–172
hazardous materials incidents, 218–220
incident priorities, 237
infrastructure emergencies, 234–236
outside fires, 185–188
rescues, 250–253
searches, 268–270
terrorism incidents, 285–286
vehicle fires, 204–206
weather-related events and national emergencies, 302–304

incidents of national consequence or interest
about, 351–354
management of, 354–355
national players, 355
news media coverage, 292

incident-specific information, 361

incident-specific understanding, 105

incinerator fires, 157

industrial processes, 170

information
and coverage sharing, 355
vs. data, 46
dissemination, 178
getting and dissemination, 91
producers and consumers of, 45
release withholding, 132

information collection briefing, 61–62

information dissemination
about emergency incident, 52
accurate and truthful, 57
breaking news stories, 80
effective and efficient way for, 99
fire cause, 345–346
and getting information, 91
media coverage, 293
media coverage success, 356–357
PIO and, 54
public information officer (PIO), 276
severe weather, 311
technologies, 95
timely, 321
vested interest in, 178

information inaccuracy, 100

information summary, 147–148

informational roles, 48–49

infrastructure emergencies
about, 233–234
hazards, inherent, 240
incident management job aids, 244–245
media coverage job aids, 245–247
media coverage of, 241–243
news reporting, 242
operational overview, 237–239
resource requirements, 241
safety considerations, 239–241
typical incidents, 234–236
typical tools, 241

ingestion, 224
inhalation, 224
injuries. *See also* civilian injuries and fatalities; firefighter injuries and fatalities
of accidental victims, 130
cause of, 255
civilian, 139
from electricity, 157, 159
to emergency responders, 290
and exposures, 130
firefighter statistics, 330
in firefighter vehicle extraction, 219
from hazardous materials, 225
information about, 148
limited, 258
minor, 331, 333
nature of, 330
from operational mechanical equipment, 162
potential, 157, 175, 192
psychological and emotional support, 130
reports on, 23
serious, 332
statistics, 185, 203, 320
from terrorism, 289
traumatic, 251, 330
and treatment urgency, 128, 253
from vehicle materials, 209
injuries and fatalities. *See also* civilian injuries and fatalities; firefighter injuries and fatalities
civilian, 225
information provider, 323
statistics on, 320
integrity, 57
International Association of Arson Investigators (IAAI), 346
international interest, 310
Internet, 2, 8, 40, 56–57, 72, 91, 99–100, 292, 310
Internet websites, 80
interpersonal roles, 48
interpersonal skills, 56, 74, 81
interviews, 85, 147
intrinsically safe equipment, 258

J

jargon, 84
job aids, incident management. *See* incident management job aids
job aids, media coverage. *See* media coverage job aids
Jobs, Steve, 95
jurisdiction, 132

L

last location determination, 271
law enforcement agencies, 237, 283
law enforcement personnel, 131, 345
life safety
of emergency responders, 189, 253
incident priorities, 19, 22, 51, 159, 172, 188, 206–208, 221, 223, 237, 241, 253, 270, 287, 290, 304, 307, 329
news media, 111, 310
of public, 270
life safety hazards, 305
lightning
vs. arson, 345
fire cause, 344
fire origin, 343
natural emergencies, 302, 307
outdoor fires, 188
at refinery, 69, 73
wildland fires, 190
liquid natural gas (LNG), 205
liquid petroleum gas (LPG), 172
liquid propane gas (LPG), 205
local news media vs. national media, 356
logistical support, 192
lost persons, 268
lost persons, identification of, 270

M

machinery rescue, 251
major accidents, 131
major emergency incidents, 68
managerial roles, 48

mass casualty incident (MCI), 128
material safety data sheet (MSDS), 220
Mayday declaration, 253, 268, 332
mechanical harm, 286
media
  access of, 35
  collaborative working with, 60
  legal rights of, 34
  working with before emergency incident, 85
media area, 62, 87, 224–225, 259, 357
media briefing, 115
media briefing book, 85
media coverage
  about, 3–4
  cooperative, 132
  demand for, 70
  dissemination of incident-related information, 293
  of outside fires, 195
  prearranged vs. unscheduled, 71–72
  stakeholder expectations, 5
media coverage job aids
  building fires, 150–153
  civilian injuries and fatalities, 326–328
  equipment fires, 166–167
  hazardous materials fires, 180–183, 198–201
  hazardous materials incidents, 228–231
  infrastructure emergencies, 245–247
  rescues, 262–265
  searches, 279–281
  terrorism incidents, 296–300
  vehicle accidents, 135–136
  vehicle fires, 214–216
  weather-related events and national emergencies, 314–318
media coverage of emergency incidents, 3–16
media incident coverage
  of arson and incendiary fires, 347–348
  of building fires, 147–150
  of civilian injuries and fatalities, 322–325
  of emergency incidents, 11, 359
  of equipment fires, 161–164
  of firefighter injuries and fatalities, 337–339
  of hazardous materials fires, 177–178
  of hazardous materials incidents, 225–226
  of infrastructure emergencies, 241–243
  of rescues, 258–259
  of searches, 275–276
  of terrorism incidents, 292–293
  of vehicle fires, 211
  of weather-related events and national emergencies, 309–311
media organizations, 65
media professionals, 4, 108
media relations and public information, 47
medical condition of victims, 253–254
medical emergencies, 309–311, 330
medical helicopters, 130–131
medical service units, emergency, 258, 289
medical surveillance, 223
microblogging, 95
mine collapse, 268
Mintzberg, Harry, 48
misleading reporting, 80
missing person(s), information about, 275
monitor roles, 48–49, 56
motor vehicle accident (MVA), 128
multiple deaths, 353
mutual aid, 116, 194, 287, 355

N

national attention and coverage, 37
National Emergency Training Center (NETC), 59
National Fallen Firefighters Foundation (NFFF), 330–331, 335
National Fire Academy (NFA), 58–59
National Fire Incident Reporting System (NFIRS), 46, 330
National Fire Protection Association (NFPA), 330
  hazardous materials incident statistics, 217

statistics on injuries and fatalities, 320
vehicle fires statistics, 204
National Incident Management System (NIMS), 15, 19, 51, 54, 106, 112–113, 309, 354, 357
national interest, 310, 354
national news media, 38
vs. local news media, 356
working with reporter, 357
national newspapers, 36
National Transportation Safety Board (NTSB), 355
natural disasters, 309–310, 352. *See also* weather-related events and national emergencies
natural fires, 344
news anchors, 4, 70, 73
news coverage
changes to, 96
elements of, 73
of emergency incidents, 4
essential role of, 41
fire impact, 163
technology in contemporary, 8
news cycle, 39
news cycle 24/7, 292
news directors, 67
news helicopters, 23, 37, 40, 67, 147, 211, 225
news media
contribution, 276
dissemination evolution, 35
focus of, 356
incidents of national consequence or interest, 292
local representatives, 355
national vs. local, 356
primary role of, 32
relationships with, 25
traditional, 92
news media coverage, success of, 33
news production activities, 69
news radio stations, 211
news reporters
and incident management process, 108
role and responsibilities, 70
role of, 1
news reporting infrastructure emergencies, 242
newspaper coverage, 36, 69
newspapers, 4
newsworthiness, 33
newsworthy incidents, 242
NFPA 704 Marking System, 219
NFPA 1006: Standard for Technical Rescuer Professional Qualifications, 253
NFPA 1035: Standard for Professional Qualifications for Fire and Life Safety Educator, Public Information Officer, and Juvenile Firesetter Intervention Specialist, 50, 56
9-1-1 communications, 107, 116
nonintervention mode, 174
nonintervention strategy, 222, 238, 240
notification of family and loved ones, 323, 334
nuclear events, 286
nursing homes, 141

O

occupancy, 141
occupancy classes, 140
offensive attack mode, 116, 144–145
offensive strategy, 222
on the record, 84
on-scene video coverage, 87
operational overviews
accidents, vehicular, 127–129
equipment fires, 158–161
hazardous materials fires, 172–175
hazardous materials incidents, 221–223
incident priorities, 143–145
infrastructure emergencies, 237–239
rescues, 253–255
searches, 270–273
terrorism incidents, 287–289
vehicle fires, 206–208
weather-related events and national emergencies, 305–306

operational tactics, 222–223
organization structure, 112, 114
outdoor activities, 249
outdoor fires, 170
outside fire safety education, 196
outside fires
- incident priorities, 188
- media coverage of, 195
- typical incidents, 185–188
- typical tools for, 194

oxygen tanks, misreporting of, 10

P

passenger vehicles, 126
passengers, 207
patient care equipment, 258
perimeter control, 273, 308
perimeter zones, 208
perimeters, 189
personal alert safety system (PASS), 146, 332
personal protective equipment (PPE), 130, 133, 162, 174, 176, 192, 209, 256, 274, 290, 307
personnel accountability, 238, 240, 256, 274
personnel accountability system, 129, 175, 189–190
photojournalists, 4–5, 15, 37, 40, 67, 85, 147, 211, 225, 259, 309, 357
physical evidence, 343
pipeline explosion and fire, coverage of, 68
policies, 109
post-incident analysis (PIA), 120, 130, 146
power lines, 306, 308
power outages, 235, 242
preparedness, 109
preplanning activities, 109
press releases, 55, 85
priorities
- vs. fire and emergency service needs, 18
- incident stabilization, 127, 329
- life safety, 127, 329
- in management, 22
- media report on, 20
- news media, 18, 110
- property conservation, 127, 329

priorities of emergency incidents, 19
priorities of incident management, 51
priorities of life safety, 110
proactive approach, 47
procedures, 109
professional collaboration, 360–362
professional development opportunities, 59
professional risks, 329
professionalism, 81
promises, 87
property conservation incident priorities, 19, 51, 110, 172, 188, 206, 221, 237, 253, 270, 287, 304, 329
protecting exposures, 176
psychological and emotional support, 130
psychological harm, 286
psychological impact, 274
public digital images, 25
public education activities, 47
public education sidebar stories/reporting, 242
public information officer (PIO)
- about, 45–46
- availability of, 22
- communication with media on firefighter injuries and fatalities, 333
- emergency incident role, 51–55
- expectations of, 54
- functions, 114
- guidance and support from incident commander, 55
- hazardous material technical experts support, 226
- incident activities, 60–62
- incidents of national consequence or interest, 293
- information dissemination, 276
- knowledge and skills, 56
- media coverage of firefighter injuries and fatalities, 333
- media expectations, 323

off-site activities, 54
positional description, 50
primary responsibilities, 50
proactive approach benefits, 46–48
proactive selection, 49
professional development of, 58–59
qualifications, 55–58
reporting inaccuracies, 24
responsibility of, 82, 115
role of, 1
selection criteria for, 55
skill development of, 59
source of accurate and credible information, 87
public information officer (PIO) role
in incident of national public and media interest, 357
in rescues, 259
and responsibilities of, 48–50
roles and activities of, 52
public information role, 242–243
public safety organizations, 93
publications, 121
push of information, 99

R

radio
broadcasting deadlines on, 37
broadcasting personalities, 69
news coverage, 36–37
radiological harm, 286
radios, 310
rail service, 234–235
rail tankers, 171
rail tracks, 240
rapid intervention teams (RITs), 146, 332
reactive approach, 47
readiness, 109
real-time footage, 23
real-time images, 310
references, 121
referrals, 84
refuge areas, 193
regional interest, 310
rehabilitation, 274
rehabilitation of personnel, 257
rehabilitation units, 275
reignition, 189
relationships. *See also* working relationships
collaborative, working with media, 10
hostile, 78
incident commander and news media, 34
negative, 80
with news media, 1, 25
relevance, 58
reporters, 4, 309. *See also* news reporters
knowledge of, 41
reputation, 177
reporter's job
about, 70–72
emergency incident repoprting preparation, 74–75
media coverage roles, 66–70
news coverage elements, 73–74
role of, 65–66
technology change impact, 72–73
reporting relationship, 86
reputation management, 9, 25, 39, 47, 49, 57, 60, 78, 80, 102–103, 333–334
rescue operations, diversity of, 253
rescue personnel, provisions for, 253
rescues. *See also* search and rescue operations
about, 249–250
incident management job aids, 260–262
incident priorities, 253
media coverage job aids, 262–265
media coverage of, 258–259
operational overview, 253–255
resource requirements, 257–258
safety considerations, 256–257
tools and equipment, 258
typical incidents, 250–253
residential buildings, 141, 320. *See also* building fires
resource management, 192
resource requirements
accidents, vehicular, 130–131
building fires, 146

equipment fires, 162–163
hazardous materials fires, 176–177
hazardous materials incidents, 224–225
infrastructure emergencies, 241
rescues, 257–258
searches, 274–275
terrorism incidents, 291
vehicle fires, 210
weather-related events and national emergencies, 309
respiratory protection, 209
Responding to Incidents of National Consequence (USFA), 354
reverse 9-1-1- systems, 93
road closures, 178, 194, 243
road detours, 243
road freight vehicles, 126
road tankers, 171
roles and responsibilities of PIO, 48
rumors, 24
rush to report, 18, 324

S

safety concerns, general, 145–146
safety considerations
building fires, 145–146
equipment fires, 158–162
hazardous materials fires, 175–176
hazardous materials incidents, 223–224
infrastructure emergencies, 239–241
rescues, 256–257
searches, 273–274
terrorism incidents, 289–290
vehicle fires, 209–210
weather-related events and national emergencies, 306–308
safety handling, 129
safety officer. *See* incident safety officer (ISO)
safety vests, 130
safety zones, 193. *See also* control zones
scene assessment and management, 273, 288
scene security, 289
scoop, 18
search and rescue, 143
search and rescue operations, 289, 305
search area, 271, 273, 275
searches
about, 264–268
challenge of operations, 267
factors of, 269
incident management job aids, 277–279
incident priorities, 270
media coverage job aids, 279–281
media coverage of, 275–276
operational overview, 270–273
perimeter, 271
resource requirements, 274–275
risks of subjects, 268
safety considerations, 273–274
typical incidents, 268–270
seasonal dangers, 27
secondary devices, 289
self-contained breathing apparatus (SCBA), 174, 190, 223
sensitivity of coverage, 319
September 11, 2001 terrorist attacks, 5–6, 285–286, 356
severe weather, 309–311. *See also* weather-related events and national emergencies
Shanksville, Pennsylvania, 6–7, 356–357
sheltering, 111, 221
sidebar coverage, 259
sidebar stories/reporting, 27, 48
situational analysis, 221
situational awareness, 288
situational environmental factors, 269–270
situational evaluation, 20
size-up
comprehensive, 305
continual, 129, 173
and forecasting, 177
importance of, 234
initial, 116–118, 127, 129, 143, 221
situation evaluation, 20
wildland fires, 189–190, 192

smart phones, 57, 98, 310
smoke inhalation, 321
social media
  communication of injuries and fatalities, 322
  emergency incidents media coverage, 310
  impact of, 100
  news and information dissemination, 80
  news organizations use of, 40
  PIO knowledge and use of, 56–57
  technology change of, 91
  technology potential, 8
  terrorism incidents coverage, 292
  threats and opportunities, 2
  use of, 99
  as valuable ally, 103
social media platforms, 72
social media usage, 24
social networking, 94
special assignment reporters, 71
specialized rescue, 250
specialized teams, 256
spokesperson roles, 48–49, 56
sprinkler systems, 145
staffed handline, 130
staging
  apparatus and equipment, 174, 176, 222, 238
  apparatus and personnel, 308
  control zones, 221, 308
  of incoming resources, 192, 207
  for media, 357
stakeholder expectations
  incident of national consequence/public interest, 354
  of incidents of firefighter injuries and fatalities, 333
  incidents of national consequence or interest, 292
  and interest, 17–18
  of media coverage, 77
  media coverage of firefighter injuries and fatalities, 333
  of news coverage, 34
  situation updates, 310
stakeholder groups, 43
standpipe systems, 142–143, 145
statistics
  building fires, 139
  firefighter injuries, 330
  hazardous materials incidents, 217
  injuries, 185, 203
  on injuries and fatalities, 320
  U.S. Fire Administration (USFA), 330
  vehicle fires, 204, 320
stories, electronic filing, 40
strategic goals, 20, 118
stress debriefing, 274
stringers, 97
structural collapse rescue, 252
structural turnout gear, 190, 192, 256, 307
structure fires, 320–321. *See also* building fires
support resources, 275
surf rescue, 251
surviving families, information for, 335
suspicious fires, 344
swiftwater rescue, 251

T

tactical objectives, 20, 118
tanker shuttles, 145
tar, 172
teamwork, 360
technical experts, 226
technical rescue, 250
technical words, 84
technology, productive use of, 56
telephone service, 236
television coverage, 37, 211
television media, 4
television news crew, 97
television personalities, 69
television reporters, 67
terminology (fire service), 106, 121
terrorism events, 284
terrorism incidents
  about, 283–285
  apparatus and equipment needed, 291

incident management job aids, 294–296
incident priorities, 287
media coverage job aids, 296–300
media coverage of, 292–293
operational overview, 287–289
resource requirements, 291
safety considerations, 289–290
targets, 286
typical incidents, 285–286
thermal burns, 321
thermal imaging cameras (TICs), 146, 161
timeliness, 81
tire fire, 172
tools, typical
building fires, 146
equipment fires, 162–163
hazardous materials fires, 177
hazardous materials incidents, 225
infrastructure emergencies, 241
outside fires, 194
rescues, 258
vehicle accidents, 131
vehicle fires, 210
tornado, 308
towing service, 210
traffic control, 129, 175, 209, 237, 274, 289, 308
traffic management, 240
traffic pattern changes, 26
traffic updates, 132
transportation. *See also* motor vehicle accident (MVA); vehicle accidents; vehicle fires
accidents, 132, 353
disruption, 225
incidents, 171
road freight vehicles, 126
road tankers, 171
transport vehicles, 126
traumatic injuries, 330
trees, downed, 308
trench rescue, 252
trucks, 126
trust, 81
tweets, 95
Twitter, 94–95, 99

U

uncertainty, 177
underground utilities, 236
undetermined fires, 344
urban search and rescue (USAR) teams, 6
urban/woodland interface, 189
U.S. Department of Homeland Security (DHS), 19, 51, 112–113, 121
U.S. Department of Transportation (DOT), 219
U.S. Fire Administration (USFA), 113, 120, 320, 330, 346
utilities, 234
utility feeds and control, 160, 208

V

vehicle accidents
emergency services needs, 250
incident priorities, 127
job aids, 134–136
media coverage of, 131–133
operational overview, 127–129
resource requirements, 130–131
tools, typical, 131
typical incidents, 125–126
vehicle extractions, 129
vehicle fires
about, 203–204
apparatus dispatched for, 210
causes of, 170, 205
incident management job aids, 212–213
incident priorities, 206
incident stabilization, 206–208
media coverage, 211
media coverage job aids, 214–216
newsworthy determination, 211
operational overview, 206–208
resource requirements, 210
safety considerations, 209–210
statistics, 320
typical incidents, 204–206
typical tools in, 210
vehicle fuel, 205
vehicle rescues, 250
vehicles, 126

ventilation techniques, 144
victims
  medical condition of, 253–254
  rescue/recovery, 255
video coverage, 73
video footage, 57
video images, 72, 147
Virginia earthquake, 92
visibility, 209

W

warm zone, 111, 174, 221
water rescue, 249, 258
water supply requirements, 144, 223
water-reactive materials, 224
weapons of mass destruction (WMD), 286
weather conditions, 190, 193, 195, 209, 302
weather events, 352
weather-related events and national emergencies
  about, 301–302
  incident management job aids, 312–314
  incident priorities, 304
  incidents, typical, 302–304
  media coverage job aids, 314–318
  media coverage of, 309–311
  operational overview, 305–306
  resource requirements, 309
  safety considerations, 306–308
websites, 100
weekly publications, 36
wildland fires
  apparatus dispatched for, 194
  contributing factors, 188–189
  firefighters protective equipment, 192
  incident management of, 190
  size-up, 190
  weather conditions, 304
wind effects, 174
windshield damage assessment, 305
winter storms, 304
witnesses, 132
working fire, 67, 116, 118
working relationships
  developing positive, 79–81
  development and maintenance of, 88
  emergency responders and counterparts, 243
  fire department and the media, 78–79
  interaction between involved fire service and media representatives, 53–54
  positive, 77, 82, 88
World Trade Center bombing, 284, 286

Y

YouTube, 8, 99, 101